General Chemistry Laboratory Manual

Third Edition

Kendra Twomey

Massasoit Community College

Kendall Hunt
publishing company

Kendall Hunt
publishing company

www.kendallhunt.com
Send all inquiries to:
4050 Westmark Drive
Dubuque, IA 52004-1840

ISBN 978-1-4652-7493-9

Contents

MASSASOIT COMMUNITY COLLEGE
CHEMISTRY LABORATORY SAFETY PROCEDURES

Laboratory safety is of the utmost importance. During each laboratory session be sure that you know the experimental procedure completely. Since any special safety precautions are discussed at the beginning of each lab period, you must be on time for the lab. You will not be allowed to enter the lab if you miss the introduction (15 minutes from the start of the lab). Also be aware of the following general safety rules; they are designed to help you have a productive and safe laboratory experience.

1. Chemical splash goggles must be worn at all times in the laboratory. Contact lenses are not recommended unless needed for therapeutic reasons. If you do wear contact lenses, please inform the instructor.

2. Loose-fitting clothing should not be worn in the laboratory. Shorts, short skirts, or dresses (legs should be covered to the knees) and open-toed shoes are not to be worn in the lab. Long hair must be tied back.

3. Know the location and proper use of the following:
    a. Eye wash stations
    b. Showers, fire blankets, and fire extinguishers
    c. Exits
    d. First Aid supplies

4. Do not pour anything down the drain without permission of your instructor. Waste chemicals are to be discarded in specially designated containers.

5. Consider all chemicals as dangerous unless you are instructed otherwise. Read labels carefully.

6. If any chemical comes in contact with your eyes, skin, or clothing, wash it off with large quantities of water and notify the instructor immediately.

7. No unauthorized experiments are to be attempted and keep the amounts of chemicals you use in an experiment consistent with what is specified.

8. Food or drink is prohibited in the laboratory.

9. Never taste a chemical. Check odors only if instructed to do so, by gently wafting some of the vapor towards your nose with your hand.

10. Gloves should be worn if a chemical can be absorbed through the skin. In any event, wash your hands thoroughly before you leave the lab.

11. Many common reagents are highly flammable and are stored in the fume hood when in the lab. Do not use them anywhere near an open flame.

12. Always turn off a Bunsen burner if you are not using it. When it is turned on, never leave it unattended.

13. If you have any allergies, list them at the bottom of page. If an experiment deals with something to which you are allergic, consult the instructor.

14. Keep your work area well organized and clean. Any broken glass should be reported to the instructor.

15. No horseplay or fooling around in the lab.

16. Report any accident to the instructor immediately.

MASSASOIT COMMUNITY COLLEGE

Chemistry Laboratory Safety Agreement

I have read and it has been explained to me and I understand and agree to follow the Massasoit Community College Chemistry Laboratory Safety Procedures listed above. Tear out and hand into your instructor.

Name (Print)________________________________ Date________________

(Signature)___________________________________

Instructor: __________________________________

Title of Course: ______________________________

Check any allergies or medical problems that your instructor should be aware of:

☐ Food/Nut ☐ Medicine/Antibiotics ☐ Other

Allergy list (be specific as possible below:

____

____

Is there any other important information that the instructor and laboratory safety officers should know about? (Examples: pregnancy, potential medical problems that could influence your participation in lab exercises, etc)

____

Lab 1
Basic Laboratory Techniques

In this experiment, you will become familiar with various types of common laboratory equipment that you will be using throughout the semester. The basic techniques to be used are learning how to use a digital balance; working with a meter stick; using a Bunsen burner; measuring volumes using beakers, Erlenmeyer flasks, and graduated cylinders; and using a digital thermometer. Units should be included with each measurement. A number without a unit of measurement is meaningless. You will also practice plotting a graph.

Equipment

100 mL graduated cylinder	100 mL beaker	Thermometer
125 mL Erlenmeyer flask	Meter stick	Striker
Bunsen burner	Gauze pad	Ring stand
Iron ring		

Experimental Procedure

Part 1 - Digital balance measurements

1. Select three items for measurement. For example, graduated cylinder, beaker, penny, etc. Write the names on the data sheet.

2. Make sure the top of the balance is clean before using. Wipe with a Kimwipe, if particles are present.

3. Press tare or re-zero button. Tare the balance before each measurement.

4. Place your item on the balance and record the mass in grams (g) to the nearest 0.01 or 0.001 g depending upon balance used. (middle column on data sheet)

5. Repeat until all three of the items have been weighed.
6. Convert the masses in grams into milligrams (mg) and kilograms (kg).

Part 2 – Measurements using a meter stick

1. Determine the smallest increment on each side of the meter stick. Record on your data sheet. When using the meter side, record measurements to the nearest 0.01 cm. Use proper significant figures for all measurements.

2. Measure and record the length (l), width (w), and height (h) of your textbook using the meter side. Calculate the volume (V).

   $V = w * l * h$

3. Repeat using the inches side of the stick. Calculate the area (A).

   $A = l * h$

4. Repeat measuring a calculator as a second object. Measure with the meter side and inches side. Find the volume (in cm^3) and area (in $inches^2$) of the calculator.

Part 3 - Temperature measurements

1. Obtain a 100 mL beaker. Fill about halfway with water. Record the temperature on the room temperature line.

2. Add some ice to the beaker of water. Stir and record the temperature. Pour water out.

3. Set up a water bath as instructed by your instructor. Place the 100 mL beaker filled about halfway with water onto the wire gauze on the iron ring. Place the Bunsen burner underneath the wire gauze. Then, attach the tubing from the Bunsen burner to the blue-topped gas spigot in the center of the bench. Turn the handle parallel with the spigot to turn on the gas. Hold the striker at an angle above the burner and strike it to obtain sparks. The Bunsen burner should ignite and produce a flame at this point. Adjust the flame by using the air vent control on the column and the gas flow control knob on the bottom of the burner.

4. Heat the water bath until boiling. Measure the temperature of the water by placing the thermometer's tip into the water. Wait for the digital reading to stabilize before recording the temperature in °C. Record the temperature in the column on the data sheet. At this point, turn the gas off by moving the handle back to the original position (perpendicular to spigot). This measurement will be time zero for your graph.

5. With the beaker of water still on the wire gauze, take the temperature of the water every 3 minutes. Stir the water periodically and take measurements with the tip in the center of the water. Record the temperature and time. Continue to take readings until you have 8 data points.

6. Graph the decrease in temperature on the graph paper. On the X-axis, plot the time and on the y-axis, plot the temperature of the water. Use as much of the graph paper as possible – spreading the graph out fully. Do not use just a small corner. Make sure to label the axis with units.

Part 4 - Volume measurements

1. Obtain a 100 mL beaker, 125 mL Erlenmeyer flask, and a 100 mL graduated cylinder. Note the increments on each piece of glassware.

2. Fill the beaker about halfway with water. Record the volume on the data sheet in mL.

3. Pour the water from the beaker into the Erlenmeyer flask. Record the volume in mL.

4. Pour the water from the Erlenmeyer flask into the graduated cylinder. Record the volume in mL.

5. Determine which piece of glassware is more accurate.

NOTE: When reading from the graduated cylinder, read volume from the bottom of the meniscus.

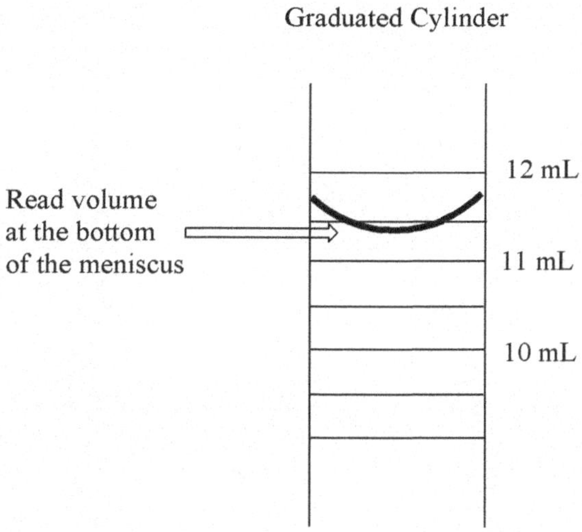

NAME: ______________________________

PARTNER: ___________________________

DATE: __________________

Basic Laboratory Techniques

Part 1 - Digital balance measurements

Item # 1: ___________________________

	milligrams	grams	kilograms
Mass	_________	__________	_________

Item # 2: ___________________________

	milligrams	grams	kilograms
Mass	_________	__________	_________

Item # 3: ___________________________

	milligrams	grams	kilograms
Mass	_________	__________	_________

SHOW CALCULATIONS.

Part 2 - Measurements using a meter stick

Smallest increment on meter side: _______________

Smallest increment on inch side: ______________

#1: Textbook

Meter side		Inches side	
Length	____________	Length	____________
Width	____________	Height	____________
Height	____________		
Volume	___________	Area	____________

2: Calculator

Meter side		Inches side	
Length	____________	Length	____________
Width	____________	Height	____________
Height	____________		
Volume	___________	Area	____________

Part 3 - Temperature measurements

	Temperature
Water from tap	__________
Ice water	__________
Boiling water	___________

Time	Temperature
0 min.	
2	
4	
6	
8	
10	
12	
14	
16	
18	

Part 4 - Volume measurements

Increments on Beaker: ______________________________

Flask: ______________________________

Cylinder: ___________________________

Volume

Beaker __________

Erlenmeyer flask ___________

Graduated Cylinder ___________

Which type of equipment is more accurate for measuring volumes? Explain your reasoning.

Lab 2
Identification of a Substance by Physical Properties

Properties are those characteristics of a substance that enable us to identify and distinguish it from other substances. Physical properties are those that can be observed without altering the composition of the substance. Physical properties such as melting point, boiling point, solubility, density, viscosity, and refractive index can be expressed quantitatively. Specific combinations of physical properties are unique to a given substance, thus making it possible to identify most substances by careful determination of several properties. This type of analysis is so important that large books have been compiled listing characteristic physical properties of most known substances.

In this experiment, two unknowns will be identified based upon their physical properties. In the first part of the experiment, an unknown solid will be determined using its density. In the second part of the experiment, an unknown liquid will be identified. The following physical properties of liquids will be used: solubility, density, and boiling point. The information obtained will be compared to a list of compounds and identified.

Equipment

4 – test tubes	2 – metal pieces	10 mL graduated cylinder	
Clamp	Hot plate	Thermometer	Boiling chip
Ring stand	250 mL beaker	Test tube holder	
25 mL graduated cylinder			

CAUTION: **1. Liquid substances are highly flammable and must be kept away from open flames.**

2. Do not pour cyclohexane waste down the drain. Dispose of all waste in the proper bottles in the fume hood.

Experimental Procedure

Part 1 – Determination of unknown solid

1. Weigh an unknown metal piece and record its mass on the data sheet. Make sure to include units.

2. Fill the 25 mL graduated cylinder about halfway with water. Record the volume of water.

3. Slowly add the metal piece into the graduated cylinder. Record the new volume of water. Determine the volume occupied by the metal by subtracting the two volumes.

4. Calculate the density of the metal piece. Remember to record density with proper significant figures.

Density = mass / volume

5. Compare the value to Table 1 and determine the type of metal.

TABLE 1	
	Density (g/cm^3)
Aluminum	2.70
Gold	19.3
Lead	11.3
Copper	8.92
Silver	10.5
Iron	7.89
Brass	8.50

6. Repeat the procedure with a second metal piece.

Part 2 – Determination of unknown liquid

1. Solubility

CAUTION: Flammable liquids

a. Record the number of your unknown liquid on the data sheet.

b. Set up three clean, dry test tubes in the test tube holder. Label tube 1 as water, tube 2 as alcohol, and tube 3 as cyclohexane.

c. Add about 5 mL of water to test tube 1, 5 mL of alcohol to test tube 2, and 5 ml of cyclohexane to test tube 3.

d. Add about 5 drops of the unknown liquid into each of the three test tubes.

e. Shake each tube by tapping on the bottom of the tube with your finger.

f. Record solubility observations (s, i, or ss). If there is an appearance of two layers, then the unknown is insoluble in that liquid (water, alcohol, or cyclohexane).

2. Density

a. Weigh a clean, dry 10 mL graduated cylinder and record its mass.

b. Carefully measure 4.0 mL of the unknown liquid into the same pre-weighed cylinder. Do not add the unknown liquid while on balance! Write down the exact volume on the data sheet.

c. Weigh the graduated cylinder with the unknown liquid. Record its mass.

d. Calculate the mass of the unknown liquid by subtracting the two mass measurements.

e. Add 1 mL of unknown to your graduated cylinder so that you have a total of 5 mL total. Weigh the graduated cylinder with the unknown liquid. Record its mass.

f. Repeat by adding another 1 mL unknown liquid to have 6 mL total.

g. Calculate the density of the unknown liquid. Remember to record density with proper significant figures.

Density = mass / volume

h. Pour the liquid into the alcohol waste bottle. Determine the average density for your unknown liquid.

3. Boiling point

a. Assemble a water bath as instructed by instructor. Place the 250 mL beaker filled about ¾ with water on the hot plate.

b. Add about 2 mL of the unknown liquid and 1 boiling chip into a clean, dry test tube.

c. Attach the test tube with the unknown liquid to the clamp on the ring stand. Move the clamp down so the depth of the liquid in the test tube is in the water bath.

d. Insert the thermometer and hold the thermometer in the test tube to a height of 1 mm to 2 mm above the liquid. (Keep cord away from hot plate surface.) Record the temperature of the vapor 1 or 2 mm above liquid when the liquid is boiling. Lift the test tube out of the water bath for 30 seconds.

e. Repeat by moving the clamp back down with the test tube into the water bath. Record the temperature once the unknown liquid is boiling again. Add more unknown liquid as needed. Repeat for the third boiling temperature. Find the average boiling point of the unknown liquid.

4. Determination of unknown liquid

a. Compare solubility, density, and boiling point values to Table 2.

*** s = soluble ss = slightly soluble i = insoluble**

TABLE 2

Substance	Density (g/mL)	Boiling point (°C)	Solubility *		
			Water	Alcohol	Cyclohexane
Acetone	0.79	56	s	s	s
2,3-butanedione	0.98	88	s	s	s
Chloroform	1.49	61	i	s	s
Cyclohexane	0.78	81	i	s	s
Ethyl propyl ether	1.37	64	s	s	s
Hexane	0.66	69	i	s	s
Isopropyl alcohol	0.79	83	s	s	s
Methyl alcohol	0.79	65	s	s	ss
Methylene chloride	1.34	40	i	s	s
Propionaldehyde	0.81	49	s	s	i

NAME: ________________________

LAB PARTNER: ____________________

DATE: _________________________

Identification of a Substance by Physical Properties

Part 1 – Determination of unknown solid

	Metal #1	Metal # 2
1. Mass of metal piece	_________	_________
2. Volume of water	_________	_________
3. New volume of water with metal	_________	_________
4. Volume of metal piece (3 – 2)	_________	_________
5. Density of metal	_________	_________
6. Type of metal	_________	_________

SHOW CALCULATIONS.

Part 2 –Determination of unknown liquid

Unknown liquid # __________

1. Solubility: soluble (s) or insoluble (i) or slightly insoluble (ss)

Water	Alcohol	Cyclohexane
_________	_________	_________

2. Density

	Trial 1	Trial 2	Trial 3
a. Mass of graduated cylinder	______	______	______
b. Volume of unknown liquid	______	______	______
c. Mass of unknown + cylinder	______	______	______
d. Mass of unknown (c – a)	______	______	______
e. Density of unknown	______	______	______
f. Average density		______________	

SHOW CALCULATIONS.

3. Boiling point
    a. Trial 1 __________
    b. Trial 2 __________
    c. Trial 3 __________
    d. Average boiling point _____________

4. Determination of unknown liquid # _____ is ____________________________

Questions

1. A sample has the density of 0.832 g/mL. What is the volume of the sample with a mass of 34.01 g?

2. An empty graduated cylinder has the mass of 55.26 g. When filled with 50 mL of an unknown liquid, it has the mass of 92.39 g. What is the density of the unknown liquid?

Lab 3
Separation of Components of a Mixture

In this experiment, three different substances in a mixture will be separated using various separation techniques. The substances are ammonium chloride (NH_4Cl), sodium chloride (NaCl), and silicon dioxide (SiO_2). The first technique called sublimation will be used to remove the ammonium chloride from the mixture. The ammonium chloride solid will be changed in its vapor leaving between the sand and salt. Extraction and decantation will be used to separate the NaCl and sand. Extraction with water will dissolve the soluble NaCl, leaving behind the insoluble sand. Decantation will then be used to separate the salt water and sand. The percentage of each component will then be determined. Ideally, the percentages at the end of the experiment should add up to 100 percent.

Separation Steps:

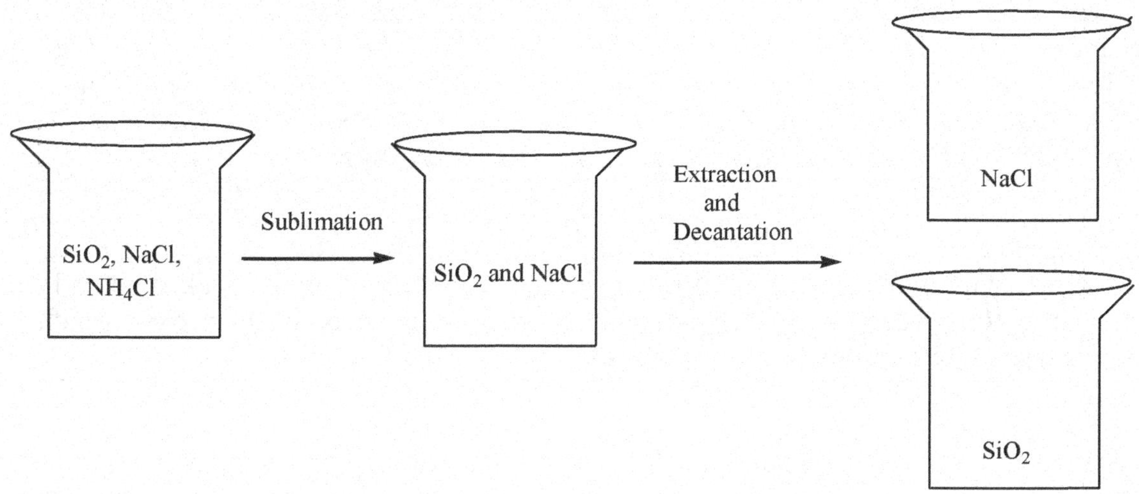

Equipment

150 mL beaker	250 mL beaker	2 -watch glasses	Stirring rod
Graduated cylinder	Striker	Wire gauze	Ring stand
Bunsen burner			

Experimental Procedure

1. Measure the mass of a clean, dry 150 mL beaker and one watch glass together. Add 0.75 to 1.00 g of the sample [ammonium chloride (NH_4Cl), sodium chloride (NaCl), and silicon dioxide (sand, SiO_2)] to the beaker. Measure and record the mass of the sample and beaker. Calculate the mass of the sample.

2. Place the 150 mL beaker and sample on a wire gauze pad on the ring stand assembly in a fume hood. Heat the sample with a Bunsen burner until white fumes cease. This process sublimes all of the ammonium chloride. Once the fumes cease, pick up the Bunsen burner and flame the sides of the beaker until the white residue is all gone. Let the beaker cool until you can hold it in your hand. Measure the mass of the beaker and record. Determine the mass of the ammonium chloride.

3. Carefully measure the mass of a dry 250 mL beaker and the second watch glass together. Add about 10 mL of distilled water to the 150 mL beaker with sample. Stir gently for 2 minutes with a stirring rod. The process extracts the soluble sodium chloride from the sand. Decant the liquid (leaving the solid behind) into the 250 mL beaker. Add another 10 mL to the 150 mL beaker, stir, and decant as before. You now have two beakers: the 150 mL beaker containing wet sand and the 250 ml beaker containing dissolved sodium chloride. MAKE SURE YOU DON'T MIX-UP THE WATCH GLASSES.

4. Place the covered 150 mL beaker containing the wet sand on the ring stand assembly. Heat slowly with a Bunsen burner stirring occasionally until the lumps break apart and the sand is dry. Remove from the heat and let cool. Measure the mass of the beaker, watch glass, and sand. Determine the mass of the sand.

5. Set the 250 mL beaker containing the NaCl solution onto the ring stand and heat gently with a Bunsen burner. Evaporate the water by gently boiling the water mixture. When the water is nearly gone, carefully place the watch glass over the beaker to catch any of the sodium chloride fragments that may splatter. When there is no more water on the underside of the watch glass and the salt is dry, remove from the heat and let cool. When cool, measure the mass of the beaker, watch glass, and sodium chloride. Determine the mass of the sodium chloride.

6. Calculate the percentage of each component in the mixture by using the following formula:

% Component = (mass of component / mass of sample) x 100

The accuracy of the experiment can be determined by determining the total percentage of the components. They should ideally add up to 100 percent. Calculate the combined total percentage recovery by adding the individual percentage of each component in the mixture.

7. Pour the sand in the wastebasket. Rinse beakers, watch glass, and stirring rod with water.

8. Repeat with second sample if time allows.

NAME: ________________________

LAB PARTNER: ___________________

DATE: _________________________

Separation of Components of a Mixture

	Trial 1	Trial 2
1. Mass of empty 150 mL beaker and watch glass	_________g	_________g
2. Mass of 150 mL beaker, watch glass, and sample	_________g	_________g
3. Mass of sample (#2 minus #1)	_________g	_________g
4. Mass of 150 mL beaker, watch glass, NaCl, and SiO_2 (after sublimation)	_________g	_________g
5. Mass of NH_4Cl component (#2 - #4)	_________g	_________g
6. Mass of empty 250 mL beaker and watch glass	_________g	_________g
7. Mass of 150 mL beaker, watch glass and dry SiO_2	_________g	_________g
8. Mass of SiO_2 component (#7 - #1)	_________g	_________g
9. Mass of 250 mL beaker, dry NaCl, and watch glass	_________g	_________g
10. Mass of NaCl component (#9 - #6)	_________g	_________g

% Component = (g component / g sample) x 100

		Trial 1	Trial 2
11. % NH_4Cl	(#5 / #3) x 100	_________	_________
12. % SiO_2	(#8 / #3) x 100	_________	_________
13. % NaCl	(#10 / #3) x 100	_________	_________
14. % Recovery (#11 + #12 + #13)		_________	_________

Questions

1. Why is it necessary to flame the top and sides the beaker during the sublimation step (step # 1)?

2. What would happen to your percentage of sand if the sand wasn't dried long enough? Explain your answer.

3. If your percentages do not add up to 100 % $\pm$ 2, what experimental errors could have occurred? Give at least two reasons.

Lab 4
Determination of Percent Sugar in Drinks

Commercial drinks contain sugar as an ingredient and in this lab you will determine the percentage of sugar in different drink samples. This will be achieved by constructing a standard percent sugar vs. density curve. Standard curves allow an individual to determine an unknown value by measuring two properties and graphing the information. For example, in the figure below, if the X value of property # 1 is determined, then a line can be drawn to the standard line to determine the value of property #2.

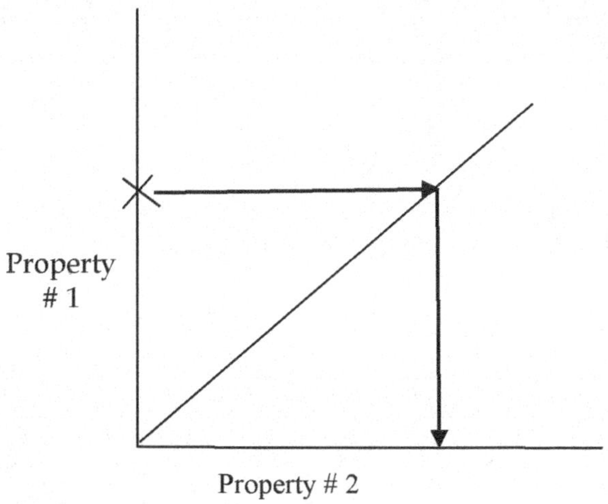

Figure 4-1. Example of a standard curve.

Equipment

25 mL graduated cylinder, Deionized water, 10 mL Volumetric pipet
Sugar, 100 mL beaker, Glass stir rod, Drinks, Pipet bulb

Experimental Procedure

1. Prepare standard sugar solutions as described: 0%, 5%, 10%, 15 % and 20% by mass. For example, a 12 % solution will have 12 g sugar and 88 g water.
    a. Measure out the required amount of sugar into the beaker. Add the required amount of water to make 100 mL of solution. Stir with a stir rod until the sugar is fully dissolved.

2. Weigh a clean, dry 25-mL graduated cylinder and record its mass.

3. Carefully measure 10 mL of the standard solution into the same pre-weighed cylinder from step 2 using the volumetric pipet and bulb. Do not add the standard solution while the cylinder is on balance. Write down the volume on the data sheet.

4. Weigh the graduated cylinder with the solution. Record its mass.

5. Calculate the mass of the sample by subtracting the two mass measurements from step 4 and 2.

6. Calculate the density of the sample. Remember to record density with proper significant figures.

$$\text{Density} = \text{mass} / \text{volume}$$

7. Determine the density for your sample solution.

8. Repeat the density determination with the drink samples using the same procedure.

9. Construct the graph using Microsoft Excel.

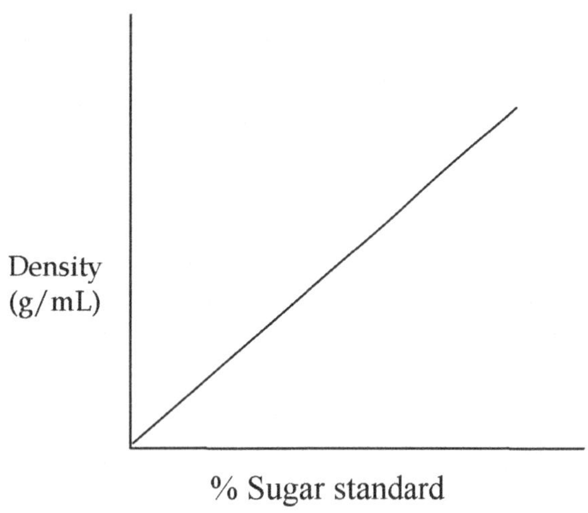

10. Determine the % sugar in your drink samples using the standard Excel graph.

NAME: ______________________

PARTNER: ___________________

DATE: ____________

Determination of percent sugar in drinks

Mass of empty 10 mL graduated cylinder: ___________

	Mass of cylinder + solution	Mass of solution	Volume of solution (mL)	Density (g/mL)
0 % sugar				
5 % sugar				
10 % sugar				
15 % sugar				
20 % sugar				

SHOW SAMPLE CALCULATION.

1. Drink # 1: ________________________

    a. Mass of empty cylinder ______

    b. Volume of sample _______

    c. Mass of cylinder + solution _______

    d. Mass of sample _______

    e Density of sample _______

SHOW SAMPLE CALCULATION.

2. Drink # 2: ________________________

    a. Mass of empty cylinder ______

    c. Volume of sample _______

    c. Mass of cylinder + solution _______

    d. Mass of sample _______

    e Density of sample _______

3. Drink # 3: ________________________

    a. Mass of empty cylinder ______

    d. Volume of sample _______

    c. Mass of cylinder + solution _______

    d. Mass of sample _______

    e Density of sample _______

4. Drink # 4: ______________________

   a. Mass of empty cylinder ______

   e. Volume of sample _______

   c. Mass of cylinder + solution _______

   d. Mass of sample _______

   e Density of sample _______

Questions

1. What is the percent sugar in the drinks based on your standard graph?

   #1. _______ # 2. _______ # 3. _______ # 4. ________

   Which type of drink contains the highest percentage of sugar?

2. Where could sources of errors occurred?

Lab 5
The Empirical Formula of an Oxide

In this experiment, the empirical formula of an oxide will be determined. The empirical formula of a compound gives the simplest ratio of the relative number of atoms in a compound. This ratio is usually a ratio of small whole numbers.

The true formula (molecular formula) gives the actual number of atoms of each element in the compound. Glucose, a sugar, has the true formula of $C_6H_{12}O_6$. The empirical formula of glucose is CH_2O. Note that the relative numbers of atoms of carbon and oxygen are the same, yet there are twice as many atoms of hydrogen relative to the number of carbon and oxygen. This is evident in both the empirical formula and the true formula.

The true formula may be found from the empirical formula if the formula mass of the compound is known. Refer to sample problem 1.

Sample Problem 1

The empirical formula for glucose is CH_2O and the formula mass for glucose is 180.06 g/mol. Find the true formula for glucose.

1. Calculate the formula mass of the empirical formula.

   $CH_2O = 30.01$ g/mol

2. Determine how many empirical formula units are in the formula mass.

   $$\frac{180.06 \text{ g/mol}}{30.01 \text{ g/mol}} = 6$$

3. Multiply the empirical formula by this integer to obtain the true formula.

   $(CH_2O)_6 = C_6H_{12}O_6$

The empirical formula of a compound may be determined experimentally by finding the mass of each of the elements that combine to form the compound. The masses are converted to moles of the element. These data can then be used to determine the simple combining ratios of the number of atoms of each element. From these mole ratios, the empirical formula can be obtained. Refer to sample problem 2.

Sample Problem 2

If 1.60 g of oxygen combines with 1.60 g of sulfur, what would be the empirical formula of the oxide of sulfur?

1. Find the number of moles of each element in the compound.

$$\text{Moles of oxygen} = \frac{1.60\text{ g}}{16.000\text{ g/mol}} = 0.0100\text{ mol}$$

$$\text{Moles of sulfur} = \frac{1.60\text{ g}}{32.06\text{ g/mol}} = 0.00499\text{ mol}$$

2. Find the ratio of the number of atoms of oxygen and sulfur by dividing by the smallest number of moles. In this case, sulfur contains the smallest number of moles, thus each element must be divided by 0.0100 the number of moles of sulfur.

$$\text{Sulfur ratio} \quad \frac{\#\text{ moles S}}{\#\text{ moles S}} = \frac{0.00499}{0.00499} = 1.00$$

$$\text{Oxygen ratio} \quad \frac{\#\text{ moles O}}{\#\text{ moles S}} = \frac{0.0100}{0.00499} = 2.00$$

3. The ratios of the number of moles indicate there are twice as many oxygen atoms as sulfur atoms in the oxide of sulfur. These numbers are the subscripts that belong with the atoms.

Empirical formula = SO_2

In this experiment, you will burn a known mass of magnesium in air. The mass of the product will be found from the mass of oxygen reacting with magnesium. By using the same type of calculation as in sample problem 2 of SO_2, the empirical formula of the oxide of magnesium will be determined.

Equipment

Magnesium ribbon	Crucible	Crucible cover	Ring stand
Clay triangle	Tongs	Striker	Wire gauze
Bunsen burner	Iron ring		

Experimental Procedure

1. Set the clay triangle onto the iron ring. Place the crucible with cover onto the clay triangle.

2. Heat the crucible and cover for 5 minutes. Using tongs set the crucible and cover on the wire gauze to cool. CAUTION: Extremely hot!

3. Allow the crucible and cover to cool to the touch. Determine the mass of the crucible and cover to the nearest 0.001 g. Record the mass on the data sheet.

4. Obtain a piece of magnesium ribbon with a mass of 0.40 g to 0.60 g. Loosely roll the magnesium into a ball. Place it in the crucible and cover. Record the mass of the covered crucible with magnesium on the data sheet.

5. Place the covered crucible with magnesium onto the clay triangle. Tilt the crucible cover over the edge slightly to allow air to get inside. (Figure 5.1)

6. Heat the crucible for approximately 5 minutes. The magnesium should be reacted when there are no signs of smoke and the contents of the crucible do not glow brightly. The product should resemble cigarette ash. Heat longer if needed. Remove and cool the crucible with cover on the wire gauze.

7. Determine the mass once the crucible and cover are cool to the touch. Record the mass on the data sheet.

8. To ensure that the magnesium has reacted completely, a second heating, known as "heating to a constant mass" should be carried out. Heat the crucible and cover again for 3 to 5 minutes. Cool the crucible and cover. Obtain the mass. This second mass should agree with 0.02 g of the first mass from step # 7. If the masses do not agree, a third heating step must be carried out, cooling, and reweighing to within 0.02 g. Record the mass on the data sheet as the final mass of crucible, product, and cover once a constant mass is obtained.

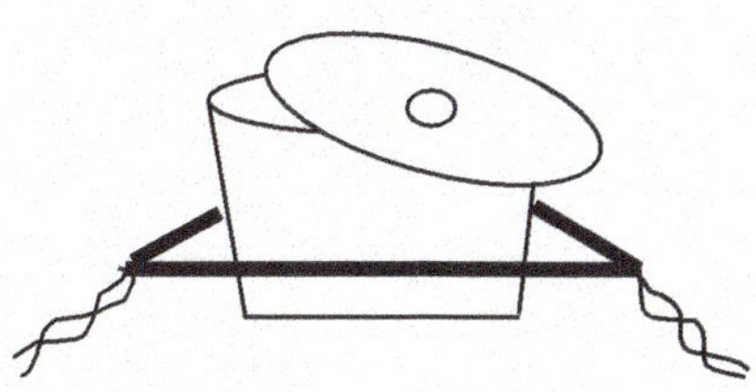

Figure 5.1 Magnesium heating setup.

Calculations

1. To find mass of magnesium, subtract line 1 from line 2.

2. To find mass of magnesium oxide, subtract line 1 from line 5.

3. To find mass of oxygen that combined with magnesium, subtract line 3 from line 6.

4. Find the moles of magnesium by dividing mass of Mg (line 3) by molar mass of Mg (24.31 g/mol).

5. Find the moles of oxygen that combined with magnesium by dividing mass of O (line 7) by molar mass of oxygen (16.00 g/mol).

6. Find the ratio of the number of moles of Mg: Divide the moles of Mg by the smallest number of moles from line 4 or 5. Find the ratio of the number of moles of O : Divide the moles of O by the smallest number of moles from line 4 or 5. Round ratios to the nearest whole number. Record ratios on the data sheet. (lines 10 and 11)

7. Find the empirical formula of the magnesium oxide using the whole number ratio of Mg (line 10) and ratio of O (line 11) as subscripts.

8. Write a balanced chemical equation for the formation of magnesium oxide from magnesium and diatomic oxygen (O_2).

Lab 6
Gravimetric Analysis of Hydrated Copper Sulfate

Gravimetric analysis involves comparing the mass of a compound before and after a chemical change has been made to the compound. Gravimetric analysis is often used in experiments to determine the concentration of a particular chemical by selective precipitation or to determine the stoichiometry of a chemical reaction. Once the identities of the participating chemical species are known, gravimetric analysis allows us to determine in what proportions they react.

Decomposition reactions involve one chemical breaking down into two or more simpler chemicals. Decomposition can occur spontaneously if the chemical is not stable, but compounds that are stable under standard conditions can be forced to decompose through the application of heat or light. Sometimes one of the products of a decomposition reaction is gaseous and can easily leave.

Some chemicals have water molecules that are very closely associated with their crystal structure. This water is not part of the actual chemical molecule, but is simply present in the crystal structure in a definite proportion. These compounds are called hydrates and the water molecules are often referred to as waters of hydration. The waters of hydration for a particular compound often have a particular stoichiometry. Usually these waters can be driven off from the compound through heating, leaving the waterless or anhydrous form of the compound.

In this experiment we will determine the stoichiometry of the waters of hydration for copper (II) sulfate. Copper (II) sulfate hydrate is blue, whereas anhydrous copper (II) sulfate is white. Anhydrous means without water. This color change provides an easily observable endpoint for the chemical change. When we heat the copper (II) sulfate hydrate, it will decompose and the waters of hydration will leave as water vapor. We can then use gravimetric analysis to determine the amount of water that was in the sample. Because we understand the stoichiometry of the reaction (see below), we can determine the number of waters of hydration using the following equation:

$$CuSO_4 \cdot nH_2O\ (s) \rightarrow CuSO_4\ (s) + nH_2O\ (g)$$

Equipment

Hydrated copper sulfate	Bunsen burner	Striker	Iron ring
Crucible	Ring stand	Wire gauze	
Tongs	Clay triangle		

Experimental Procedure

1. Measure the mass of a crucible and record on your data sheet.

2. Weigh out 5 g to 7 g of copper (II) sulfate hydrate into the crucible; record the mass.

3. Setup a heating apparatus as instructed. Place the crucible with copper (II) sulfate hydrate onto a clay triangle.

4. Heat the crucible using the Bunsen burner for 15 to 20 minutes, or until the sample has turned completely white.

5. Remove the crucible from the heat using tongs and allow it to cool completely on wire gauze.

6. Record the mass of the crucible and anhydrous copper (II) sulfate.

7. Repeat steps 4 through 6 heating 5 to 10 minutes until the mass of the crucible and anhydrous copper (II) sulfate remains constant within $\pm$ 0.02 g.

8. Repeat with a second sample, if time permits.

NAME: ________________________

LAB PARTNER: ___________________

DATE: _________________________

Gravimetric Analysis of Hydrated Copper Sulfate

	Trial 1	Trial 2
1. Mass of crucible	________	________
2. Mass of crucible and copper (II) sulfate hydrate	________	________
3. Mass of copper (II) sulfate hydrate	________	________
4. Mass of crucible and copper sulfate after first heating	________	________
5. Mass of crucible and copper sulfate after second heating	________	________
6. Mass of crucible and copper sulfate after third heating (if necessary)	________	________
7. Mass of anhydrous copper (II) sulfate after final heating (line 5 or 6 minus line 1)	________	________
8. Moles of anhydrous copper (II) sulfate	________	________
9. Mass of water (line 2 minus line 7)	________	________
10. Moles of water	________	________
11. Ratio of (moles of water) / (moles of $CuSO_4$)	________	________
12. Formula of copper (II) sulfate hydrate	________	________

SHOW CALCULATIONS.

Questions

1. What would happen to the hydrate formula if the sample wasn't dried to a constant mass?

2. Calculate the formula of $FeCl_3 \cdot nH_2O$ if you started with 8.99 g of iron (III) chloride hydrate. After heating, the mass was reduced to 4.5 g of anhydrous iron (III) chloride.

Lab 7
Separation of Ions by Chromatography

Chromatography is a technique used to separate and identify individual components in a mixture. The name chromatography arose from the separation of colored pigments; it literally means 'color writing'. In its simplest form, a chromatographic set-up consists of an immobilized substance (called the "stationary phase") over which a gas or a liquid (the "mobile phase") moves. Components of a mixture injected into the mobile phase can be separated based on their tendency to move along with the mobile phase rather than be attracted to the stationary phase. A variety of chromatographic techniques have been developed. The most commonly used ones are, thin layer chromatography (TLC), paper chromatography, gas chromatography (GC), and high performance liquid chromatography (HPLC).

Thin-layer chromatography (TLC) is one of the easiest chromatographic techniques. A thin layer of a suitable solid substance is coated on a sheet of glass (which is the stationary phase). By immersing one edge of the sheet in an appropriate solvent (the mobile phase), the solvent is drawn up the sheet by capillary action, and the compounds of interest are carried along at different rates. This is commonly called "developing" the plate.

The components that are adsorbed most strongly to the plate move slowly up the plate and vice versa. The solubility of the components in the mobile phase also affects the separation. If one component is better soluble in the mobile phase, it moves faster than another that is less soluble. The different rates will produce the desired separation. A chromatogram is produced showing the movement of the components up the filter paper.

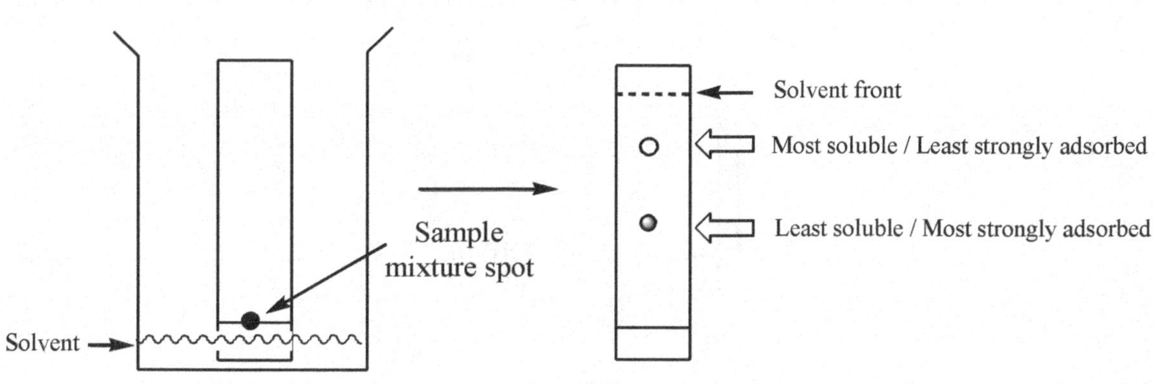

Chromatogram

The movement of the component is characterized by its retention factor, R_f. It is specific towards the solvent system. Retention factors will change upon changing the solvents system.

$$R_f = \frac{\text{Distance traveled by spot from point of origin}}{\text{Distance traveled by solvent}}$$

The origin is where the original sample mixture was spotted on the TLC plate. A line in pencil is usually drawn to identify it. The distance traveled by solvent is measured from the point of spotting and NOT from the edge of the TLC plate. To measure the component distances, measure from the origin line to the middle of spot. The components in a mixture can be identified by their R_f values.

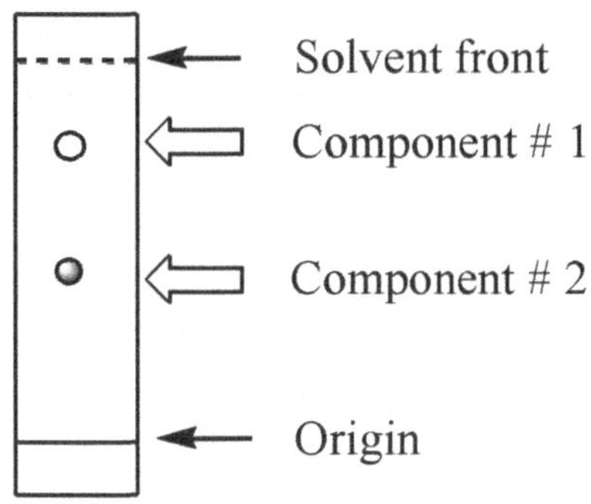

Sample Problem.

Calculate the R_f values for the two components in the chromatogram shown below. The distance traveled by the solvent front is 12.5 cm, by component # 1 is 9.2 cm, and by component # 2 is 5.5 cm.

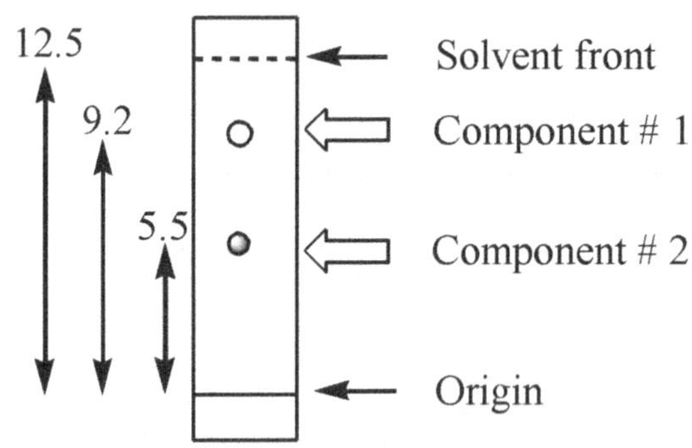

Component # 1: R_f = 9.2 cm / 12.5 cm = 0.74

Component # 2: R_f = 5.5 cm / 12.5 cm = 0.44

In this experiment, you will separate a mixture of metal ions and separate the colored dyes in markers. In the first part, you have a mixture of Fe^{3+}, Cu^{2+}, and Co^{2+} ions. Solutions of the ions by themselves will be run next to mixtures of the ions. Based upon their R_f values, you will identify which ions are present in the mixtures. In the second part, markers (green, black, and red) will be spotted on the paper, and separated into the dyes of which they are composed.

Equipment

3 - markers	2 - 400 mL beakers	Stirring rod	Parafilm
Stapler	Filter paper		

Experimental Procedure

Part 1 - Separation of metals ions from a mixture

1. Cut a strip of filter paper that will fit just inside the 400 mL beaker and slightly longer than the height of the beaker.

2. Drape the filter paper over the stirring rod. Place the stirring rod with paper into the beaker. Staple the top of filter paper together once the bottom of the filter paper is touching the bottom of the beaker. (Figure 7-1) Remove paper from beaker.

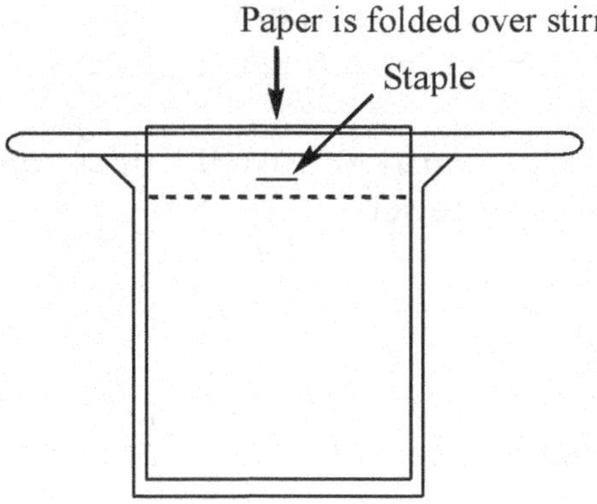

Figure 7-1. Setup for chromatography

3. Draw a line about 1 cm to 2 cm from the bottom of the filter paper in <u>pencil</u>.

4. Spot the ions (Fe^{3+}, Cu^{2+}, and Co^{2+}, and the two unknown mixtures) using small glass capillary tubes onto the pencil line by dipping the capillary tube in the ion solution and touch the paper quickly to spot ion on the pencil line.

5. Allow the spots to dry.

6. Pour a small amount of the solvent into the 50 mL beaker. The solvent is a mixture of HCl, acetone, and water (1:8:1). While holding the spotted paper on the outside of the 400 mL beaker, pour the solvent into the 400 mL beaker until it is just below the origin line. You do not want the solvent to be above the pencil line before starting. Bring it to your bench area before inserting.

7. Set the stirring rod with paper folded over into beaker carefully. (Figure 7-2)

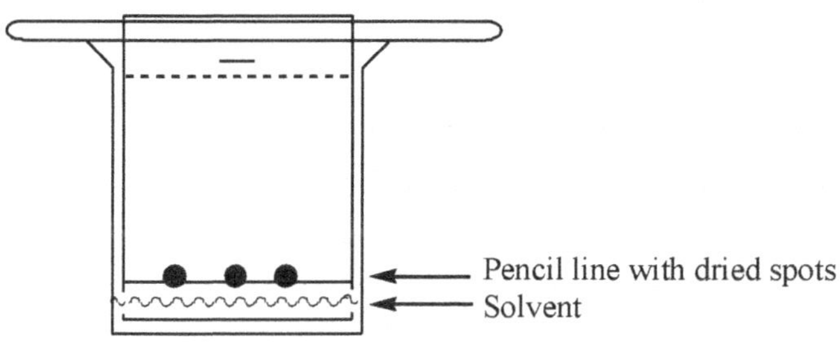

Figure 7-2. Setup with solvent in beaker.

8. Cover the beaker with Parafilm.

9. Allow the solvent to travel about ¾ of the way up the filter paper. Be careful not to move the beaker during this process

10. Once the solvent has moved ¾, remove the filter paper from the beaker and allow it to dry on a paper towel.

11. Circle any visible colored spots with a pencil.

12. Using the spray bottle, while holding the paper over a sink, spray the solution of potassium hexacyanoferrate over the paper. Place on a paper towel.

13. Calculate the retention factors for the known ions.

14. Identify the ions in the unknowns based upon the known ions R_f values.

15. Sketch the pattern of ion separation onto the data sheet. Also, staple the original chromatogram to either your data sheet or your lab partner's data sheet.

Part 2 - Separation of colored dyes in white board markers

1. Cut a strip of filter that will fit just inside the 400 mL beaker and slightly longer than the height of the beaker.

2. Drape the filter paper over the stirring rod. Place the stirring rod with paper into the beaker. Staple the top of filter paper together once the bottom of the filter paper is touching the bottom of the beaker. (Figure 7-3) Remove paper from beaker.

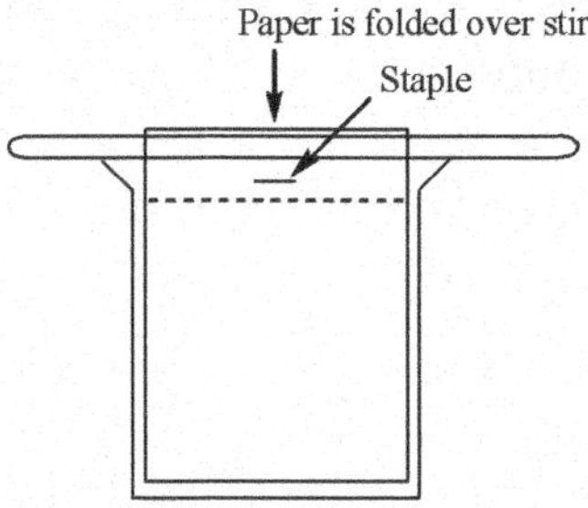

Figure 7-3. Setup for chromatography.

3. Draw a line about 1 cm to 2 cm from the bottom of the filter paper in pencil.

4. Spot the markers on the pencil line.

5. The solvent is a mixture of alcohol and water (1:1). While holding the spotted paper on the outside of the 400 mL beaker, pour the solvent into the 400 mL beaker until it is just below the origin line. You do not want the solvent to be above the pencil line before starting. Bring back to bench area before inserting.

6. Set the stirring rod with paper folded over into beaker carefully. (Figure 7-4)

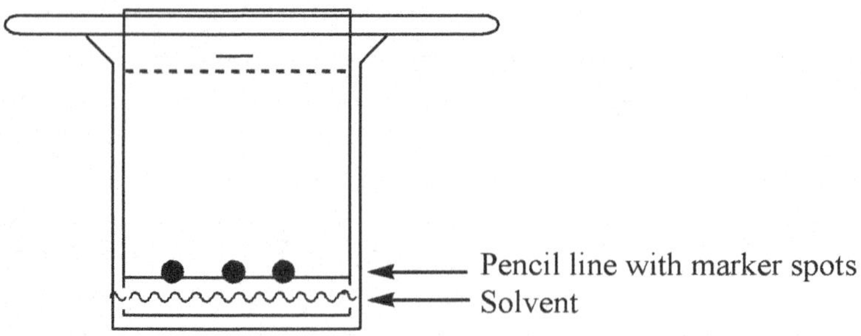

Figure 7-4. Setup with solvent in beaker.

7. Cover the beaker with Parafilm.

8. Allow the solvent to travel about ¾ of the way up the filter paper. Be careful not to move the beaker during this process.

9. Once the solvent has moved ¾, remove the filter paper from the beaker and allow it to dry on a paper towel.

10. Sketch the pattern of color separation onto the data sheet. Use arrows to label the color separation of the dyes.

NAME: ______________________

LAB PARTNER: __________________

DATE: _______________________

Separation of Mixtures by Chromatography

Part 1 - Separation of metals ions from a mixture

Solvent front distance _________________

Sample	Distance Ion Traveled	Retention Factor
Cu^{2+}		
Fe^{3+}		
Co^{2+}		

	Distance Spot 1 Traveled	Distance Spot 2 Traveled
Unknown # 1		
Unknown # 2		

	Retention Factor - Spot 1	Retention Factor - Spot 2
Unknown # 1		
Unknown # 2		

Ions in Unknowns*	Cu^{2+}	Fe^{3+}	Co^{2+}
Unknown # 1			
Unknown # 2			

* Place a check if ion was present in the unknown.

Chromatogram of Ion Separation

Part 2 - Separation of colored dyes in white board markers

Color Separation

Questions

1. What would happen to your chromatogram if you moved the beaker while the solvent was moving up the paper?

2. Why do you think it was important to use a very small amount of sample when spotting the plate?

3. Do you expect that changing the solvent will change the R_f values for a given compound? Explain your reasoning

4. If two components have identical R_f values, does this mean they necessarily have the same structure? Explain why or why not.

5. Why did the origin line have to be drawn in pencil (not pen)?

Lab 8
Chromatography of Analgesic drugs

Chromatography is a technique used to separate and identify individual components in a mixture. The name chromatography arose from the separation of colored pigments; it literally means 'color writing'. In its simplest form, a chromatographic set-up consists of an immobilized substance (called the "stationary phase") over which a gas or a liquid (the "mobile phase") moves. Components of a mixture injected into the mobile phase can be separated based on their tendency to move along with the mobile phase rather than be attracted to the stationary phase. A variety of chromatographic techniques have been developed. The most commonly used ones are, Thin Layer Chromatography (TLC), Paper Chromatography, Gas Chromatography (GC), and High Performance Liquid Chromatography (HPLC).

Thin-Layer chromatography (TLC) is one of the easiest chromatographic techniques. A thin layer of a suitable solid substance is coated on a sheet of glass (which is the stationary phase). By immersing one edge of the sheet in an appropriate solvent (the mobile phase), the solvent is drawn up the sheet by capillary action, and the compounds of interest are carried along at different rates. This is commonly called "developing" the plate.

The components that are adsorbed most strongly to the plate move slowly up the plate and vice versa. The solubility of the components in the mobile phase also affects the separation. If one component is better soluble in the mobile phase, it moves faster than another that is less soluble. The different rates will produce the desired separation. A chromatogram is produced showing the movement of the components up the filter paper.

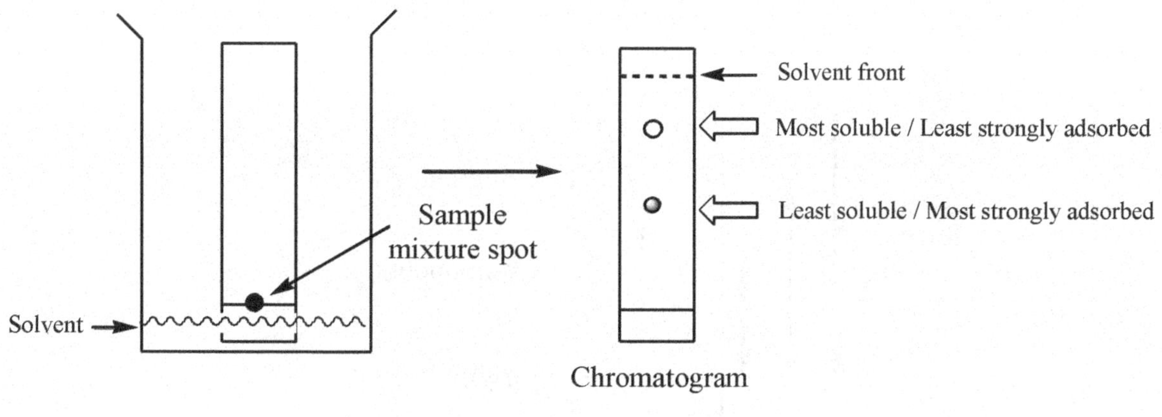

An appropriate detector (either a UV detector or a suitable staining liquid) is used to observe where the components traveled on the plate.

The movement of the component is characterized by its retention factor, R_f. It is specific towards the solvent system. Retention factors will change upon changing the solvents system.

$$R_f = \frac{\text{Distance traveled by spot from point of origin}}{\text{Distance traveled by solvent}}$$

The origin is where the original sample mixture was spotted on the TLC plate. A line in pencil is usually drawn to identify it. The distance traveled by solvent is measured from the point of spotting and NOT from the edge of the TLC plate. To measure the component distances, measure from the origin line to the middle of spot. The components in a mixture can be identified by their R_f values.

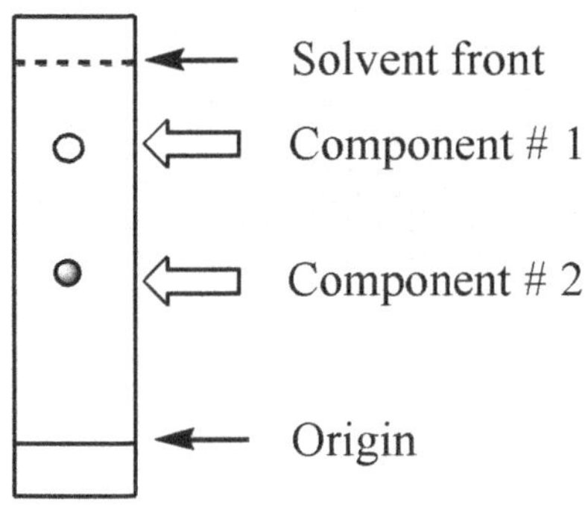

Sample Problem.

Calculate the R_f values for the two components in the chromatogram shown below. The distance traveled by the solvent front is 12.5 cm, by component # 1 is 9.2 cm, and by component # 2 is 5.5 cm.

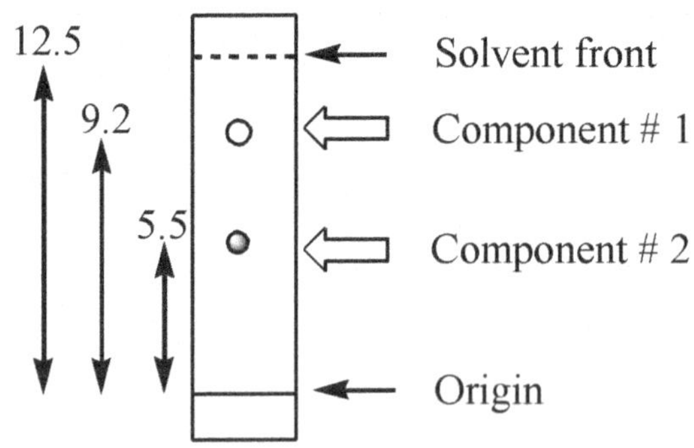

Component # 1: R_f = 9.2 cm / 12.5 cm = 0.74

Component # 2: R_f = 5.5 cm / 12.5 cm = 0.44

In this experiment, you will spot individual samples of Aspirin, Tylenol, Ibuprofen, Caffeine and an unknown drug mixture on the TLC plate. Based on their R_f values, you will identify what compounds are present in the drug mixture.

Equipment

2- 400 mL beaker	TLC Plate	Parafilm	capillary tube	Filter paper
2 - Stirring rod	Filter paper			

Experimental Procedure

1. Obtain a TLC plate and about 1 cm from the bottom of the TLC plate mark a spot in ***pencil.***

2. Spot the different solutions of drug samples using a small glass capillary tube onto the pencil spots. Dip the capillary tube in the solution and touch the plate quickly to spot the solution on the pencil mark. Make sure that all the spots are separate from each other and are on the same horizontal line.

3. Do not mix up the capillary tubes. The solutions of drugs are fairly concentrated and hence it is sufficient to spot just once on the TLC plate.

4. Allow the spots to dry.

5. Pour about 6-7 mL of the solvent into the 50 mL beaker. The solvent is a mixture ethyl acetate, ethanol and acetic acid (25:1:1). Pour the solvent into the 400 mL beaker.

6. **Place the TLC plate gently into the beaker making sure that the solvent level is touching the plate but is not above the spots**. The figure 8-1 below illustrates this.

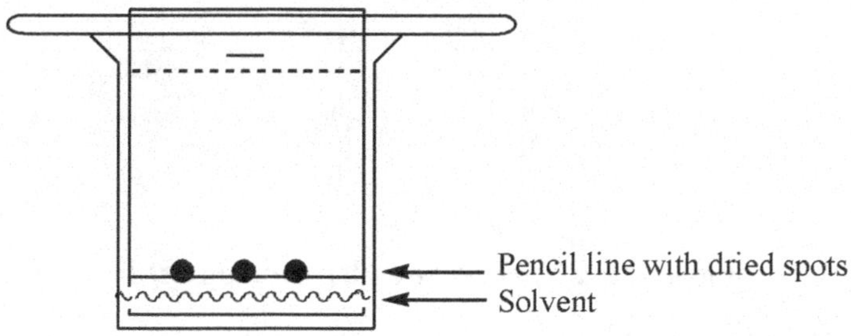

Figure 8-1. Setup with solvent in beaker.

7. Cover the beaker with Parafilm.

8. Allow the solvent to travel about ¾ of the way up the plate. Be careful not to move the beaker during this process

9. Once the solvent has moved ¾, remove the TLC from the beaker and **quickly mark the solvent level.** Allow the plate to dry on a paper towel.

10. Place the plate under the UV lamp (254 nm) to view where the spots traveled on the plate. While observing under the UV lamp, draw a light pencil line around each spot.

11. Calculate the retention factors (R_f) for the known drug samples.

12. Identify the components present in the unknown drug mixture based on the R_f values of known drug samples.

13. Sketch the pattern of component separation in the space below.

Part 2 - Separation of colored dyes in white board markers and black pens

1. Cut a strip of filter that will fit just inside the 400 mL beaker and slightly longer than the height of the beaker.

2. Drape the filter paper over the stirring rod. Place the stirring rod with paper into the beaker. Staple the top of filter paper together once the bottom of the filter paper is touching the bottom of the beaker. (Figure 8-2) Remove paper from beaker.

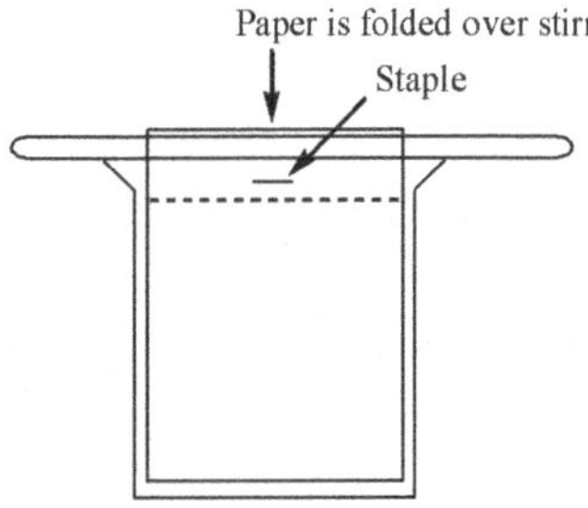

Figure 8-2. Setup for chromatography.

3. Draw a line about 1 cm to 2 cm from the bottom of the filter paper in pencil.

4. Spot the markers on the pencil line.

5. Pour a small amount of the solvent into the 50 mL beaker. The solvent is a mixture of alcohol and water (1:1). While holding the spotted paper on the outside of the 400 mL beaker, pour the solvent into the 400 mL beaker until it is just below the origin line. You do not want the solvent to be above the pencil line before starting.

6. Set the stirring rod with paper folded over into beaker carefully. (Figure 8-3)

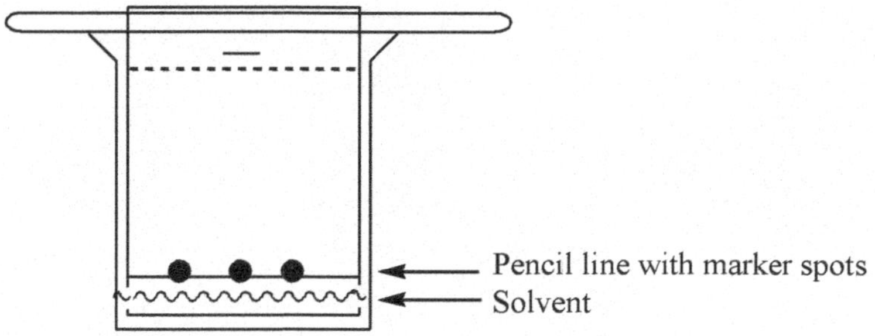

Figure 8-3. Setup with solvent in beaker.

7. Cover the beaker with Parafilm.

8. Allow the solvent to travel about ¾ of the way up the filter paper. Be careful not to move the beaker during this process.

9. Once the solvent has moved ¾, remove the filter paper from the beaker and allow it to dry on a paper towel. Sketch the pattern of color separation onto the data sheet. Use arrows to label the color separation of the dyes.

NAME: ______________________________

PARTNER: ___________________________

DATE: __________________

Chromatography of Analgesic drugs

Qualitative Interpretation of your TLC plate

1. Did the three pure analgesics compounds and caffeine each produce a single spot? Did the spots move different distances?

Calculation of R_f Values

Distance traveled by Solvent (from the point of spotting)

= __________________

Sample	Distance Traveled by spot	Retention Factor (R_f)
Aspirin		
Tylenol		
Ibuprofen		
Caffeine		

Unknown Drug Mixture

Spots	Distance traveled by spot	Retention Factor (R_f)	Component present
Spot 1			
Spot 2			
Spot 3			

The Drug mixture contains ____

Part 2 - Separation of colored dyes in white board markers

Color Separation

Questions

1. Suggest possible advantages and disadvantages of using a longer (taller) TLC plate

2. What would happen to your chromatogram if you moved the beaker while the solvent was moving up the paper?

3. Why do you think it was important to use a very small amount of sample when spotting the plate?

4. Do you expect that changing the solvent will change the R_f values for a given compound? Explain your reasoning

5. If two components have identical R_f values, does this mean they necessarily have the same structure? Explain why or why not.

6. Why should you use pencil to mark the spots (as opposed to using a pen)?

Lab 9
Determination of $KClO_3$ in a Mixture

In experiment 3 you separated a mixture of sand and salt based on their differences in solubility in water. But what would you have done if they both were water soluble? Chemically reacting one of the compounds and not the other is one way to do this. In this experiment you will be given a mixture of potassium chlorate and sodium chloride, both are water soluble. Your assignment is to determine the percent by mass of potassium chlorate in the mixture. To do this you will exploit the fact that potassium chlorate decomposes when heated while sodium chloride does not. Your instructor will show you how to arrange the set up for the experiment. A weighed sample of the mixture will be heated in a clean test tube. The sample as well as the test tube must be very clean because the potassium chlorate will produce oxygen gas when it is heated. This can result in an explosion if combustible materials are mixed with the $KClO_3$.

The reaction is:

$$2KClO_3 + NaCl \rightarrow 2KCl + NaCl + 3O_2\uparrow$$

The difference in mass before and after the reaction will be equal to the mass of O_2 produced. The moles of O_2 will be calculated and used, with the balanced equation, to find the moles of $KClO_3$ that must have been present in the mixture. For every 2 moles of $KClO_3$ that react, 3 moles of O_2 are produced. The moles of $KClO_3$ will be changed to grams by using its molar mass, and this will be used to find the percent by mass of $KClO_3$ in the mixture. The mass and moles of NaCl will also be found and from this a mole percent will be calculated.

$$\text{mass \%} = \frac{\text{mass of compound (100)}}{\text{mass of mixture}} \qquad \text{mole \%} = \frac{\text{moles of A (100)}}{\text{(moles A+ moles B)}}$$

Equipment

Test tube Bunsen burner Ring stand Clamp 2.5g of the mixture

Experiment procedure

1. Weigh a clean, dry test tube to 0.001g.

2. Add about 2.5 grams of the mixture to the test tube and weigh. Attach the test tube with a clamp to a ring stand. Keep the test tube parallel to the table top.

3. Heat the test tube until the sample melts. Only heat the part of the test tube containing the mixture. Once it melts, the $KClO_3$ will decompose and oxygen gas will bubble out of the test tube. Continue heating for about 5 minutes, rotating the test tube by hand, until the reaction appears over.

4. Let the test tube cool and weigh. Reheat for 5 minutes, let cool and reweigh. Continue this step until a constant weight is attained.

5. Dispose of the contents of the test tube as instructed.

Calculations:

8. #7 x 1mol/32g

9. #8 x 2mol KClO3/3mol O2

10. #9 x 122.6g/mol

12. #11 x 1mol/58.5g

13. #10 (100)/#3

14. #9 (100)/(#9 + #12)

Lab 10
Precipitation Reactions

In a general sense, solubility can be thought of as the tendency of a substance (solute) to dissolve in another substance (solvent). For qualitative purposes, such terms as "soluble", "insoluble", and "slightly or partially soluble" can be used to describe these tendencies. Chemistry texts contain solubility rules which show the tendencies of several ionic compounds to dissociate in water.

Ionic compounds dissociate in water rather than dissolve. In this process, the crystal lattice of the solid is attacked by water molecules. Because water molecules contain partial positive and negative charges, they are attracted to ions. If this attraction is stronger than the attractions holding the crystal together, then the crystal slowly dissolves producing ions that freely move throughout the solution. Each ion becomes surrounded by, and bonded by water molecules which prevent the ions from reforming the crystal. This process is called hydration. If the ionic attractions are stronger than those between water and the ions, then dissociating does not happen and the ionic substance is termed insoluble. An example of a dissociation equation is:

$$Na_2SO_4(s) \rightarrow 2Na^+(aq) + SO_4^{-2}(aq)$$

If two aqueous ionic solutions are mixed, a precipitate may form. The mixture will look cloudy and may be colored. The precipitate is a product of a double replacement reaction that has a low solubility in water.

In this experiment different ionic solutions will be mixed and the results observed. Molecular, ionic and net ionic equations will be written for those reactions that showed the formation of precipitates,. The following example may be helpful. A mixture of sodium chloride and silver nitrate solutions results in the formation of a precipitate. We find the products by switching the names of the reactants:

sodium chloride + silver nitrate → sodium nitrate + silver chloride

By applying the solubility rules we find that silver chloride is insoluble, so it is the precipitate.

The molecular equation is: $NaCl(aq) + AgNO_3(aq) \rightarrow NaNO_3(aq) + AgCl\downarrow$
The ionic equation is: $Na^+(aq) + Cl^-(aq) + Ag^+(aq) + NO_3^-(aq) \rightarrow Na^+(aq) + NO_3^-(aq) + AgCl\downarrow$
The net ionic equation is: $Ag^+(aq) + Cl^-(aq) \rightarrow AgCl\downarrow$

Equipment

4 small test tubes or spot plate test tube rack
Solution Set:

0.1M $AgNO_3$	0.1 M $BaCl_2$	0.1 M Na_2CO_3
0.1M $(NH_4)_2SO_4$	0.1M $Ca(NO_3)_2$	0.1M K_3PO_4

Experimental Procedure

1. Obtain any two of the solutions from the set above, and a clean plastic spot plate.

2. Place a few drops of each solution into a depression on the spot plate. The mixture will look cloudy instantly if a precipitate forms, or stay clear if no reaction occurs. Write "ppt" when a combination produces a precipitate. Write "NR" for no reaction when no precipitate forms.

3. Repeat step 2 until all possible combinations have been done. Rinse off the spot plate when finished.

Calculations:

1. Write molecular equations for each reaction that showed a precipitate being formed.

2. Write ionic equations for each reaction that showed a precipitate being formed.

3. Write net ionic equations for each reaction that showed a precipitate being formed.

Lab 11
Titration of Acids and Bases: Standardization of Sodium Hydroxide and Determination of Percent Acetic Acid in Commercial Vinegar

One of the most common and familiar reactions in chemistry is the reaction of an acid with a base. This reaction is termed neutralization. The products of a neutralization reaction are always a salt and water. The cation of the salt always is from the base and the anion of the salt comes from the acid.

$$\underset{\text{Acid}}{HCl} + \underset{\text{Base}}{NaOH} \rightarrow \underset{\text{Salt}}{NaCl} + \underset{\text{Water}}{H_2O}$$

In this experiment you will use a neutralization reaction to accurately determine the concentration of a sodium hydroxide solution. The process of determining the concentration of the solution is called standardization. The technique of accurately measuring the volume of a solution required to react with another reagent is termed titration. You will be using a burette as the equipment for performing the titrations.

An indicator solution is used to determine when an acid has exactly neutralized a base or vice versa. A suitable indicator changes color when equivalent amounts of acid and base are present in the solution. The color change is termed the end point of the titration. Indicators change colors due to different pH values. Phenolphthalein, for example, changes from colorless in an acidic solution ($pH < 7$) to pink in a basic solution ($pH > 7$). By changing the pH of the solution, phenolphthalein's structure is changed slightly (See below). The structure will change back when the pH is readjusted.

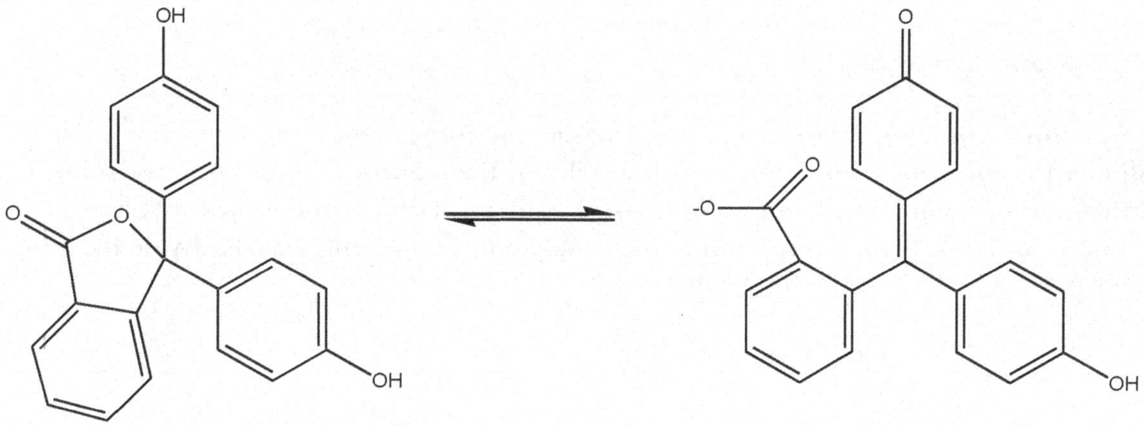

Acidic Form — **Colorless**

Basic Form — **Pink**

By definition, one gram equivalent weight of an acid is that mass of the acid in grams that will provide 1 mole of protons in a reaction. One gram equivalent weight of a base is that mass of the base in grams that will provide 1 mole of hydroxide ions in a reaction or that will react with 1 mole of protons. A gram equivalent weight is often referred to an equivalent.

Or more simply put,

$$\text{Equivalent of an acid} = \frac{\text{FW of the acid}}{\text{\# of H+}}$$

$$\text{Equivalent of an base} = \frac{\text{FW of the base}}{\text{\# of OH-}}$$

Sample calculation for equivalent weight

1 mole of HCl: $1 \text{ equivalent of HCl} = \frac{36.45 \text{ g}}{1 \text{ H+}} = 36.45 \text{ g equivalent weight}$

1 mole of H_2SO: $1 \text{ equivalent of } H_2SO_4 = \frac{98.08 \text{ g}}{2 \text{ H+}} = 49.04 \text{ g equivalent weight}$

1 mole of H_3PO_4: $1 \text{ equivalent of } H_3PO_4 = \frac{98.00 \text{ g}}{3 \text{ H+}} = 32.67 \text{ g equivalent weight}$

HCl is a monoprotic acid containing only one proton, whereas H_2SO_4 is diprotic so we must divide the formula mass by two. H_3PO_4 is triprotic with three available protons.

Normality (N) is the most convenient method for expressing concentrations for volumetric work with acids and bases. Normality is the number of equivalents of solute per liters of solution. Often, one will not use 1 equivalent of an acid or base, but a portion of an equivalent. To determine how many equivalents are present, divide the grams present by the equivalent weight.

$$\text{Normality} = \frac{\text{equivalents of solute}}{\text{L of solution}}$$

$$\text{\# of equivalents of solute} = \frac{\text{g present}}{\text{equivalent weight}}$$

Sample problem 1.

Determine the normality of an NaOH solution if you dissolve 0.450 g of NaOH in 95.2 mL of water.

$$\text{Equivalent weight of NaOH} = \frac{40.00 \text{ g}}{1 \text{ OH-}} = 40.00 \text{ g}$$

$$\text{\# of equivalents of NaOH} = \frac{\text{g present}}{\text{equivalent weight}} = \frac{0.450 \text{ g NaOH}}{40.00 \text{ g NaOH}}$$

$$= 0.0113 \text{ equivalents of NaOH}$$

$$\text{Normality of NaOH} = \frac{\text{equivalents of NaOH}}{\text{L of NaOH solution}} = \frac{0.0113 \text{ equivalents}}{0.0952 \text{ L}}$$

Normality of NaOH = 0.119 N

Relationship between normality and molarity

Equivalent weight and formula weight are related; therefore, normality and molarity are related. Normality (N) equals the number moles multiplied by molarity (M). Normality is greater than or equal to molarity.

$N = a \times M$ where a = # of moles H^+ or OH^- ions

Sample problem 2.

a. Convert 1 N HCl to molarity of HCl.

$M = N / a = 1 \text{ N} / 1 \text{ mole } H^+ = 1M \text{ HCl}$

b. Convert 1N H_2SO_4 to molarity of H_2SO_4.

$M = N / a = 1 \text{ N} / 2 H^+ = 0.5 \text{ M } H_2SO_4$

In this experiment, a solution of NaOH will be standardized, and then used to determine the percent acetic acid in vinegar. For the standardization experiment, a titration will be performed against potassium acid phthalate ($KHC_8H_4O_4$, abbrev. PAT). Its structure is shown below.

PAT =

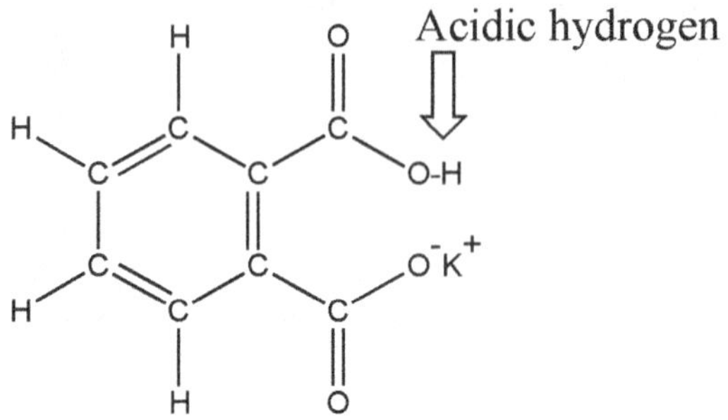

PAT is known as a standard acid material. It has been chosen as a standardizing agent for acid-base titrations for the following reasons: first: the solid material can be obtained in a very pure state (99.99% plus pure); second: it is inexpensive; and third: it is a very stable chemical compound. It is a monoprotic acid with an acidic hydrogen bonded to oxygen. Its formula mass is 204.2 g. Therefore, 1 equivalent is 204.2 g.

In the titration of NaOH against PAT, an equal number of equivalents of acid and base are present at the end point of the titration. In other words, at the end point, equivalents acid = equivalents base. This may be observed from the stoichiometry of the balance chemical equation. The PAT will be used to find the molarity of the NaOH solution (See sample problem 3).

$$NaOH + PAT \rightarrow H_2O + PAT\text{-salt}$$

Sample problem 3.

What is the normality of a sodium hydroxide solution if 35.75 mL of the solution is required to neutralize 1.070 g of PAT?

Equivalents PAT = Equivalents NaOH at the end point of the titration

$$\text{Normality} = \frac{\dfrac{1.070 \text{ g PAT}}{204.2 \text{ g / equivalent}}}{0.03575 \text{ L}} = 0.147 \text{ N NaOH}$$

A dilution will decrease the concentration of the solute in a solution. A high concentration stock solution can be used to make a lower concentration solution. A volumetric flask is used for making the exact concentrations needed. The flask has a mark that allows users to know where to add solvent to. The equation to calculate the amount of high concentration solution needed to make the lower concentration is

$$M_1 \text{ x } V_1 = M_2 \text{ x } V_2$$

where M is molarity and V is volume.

Sample problem 4.

Calculate the volume needed of a 2.5 M HCl solution to make a 500 mL of a 0.47 M HCl solution.

$$M_1 \text{ x } V_1 = M_2 \text{ x } V_2$$
$$2.5 \text{ M x } V_1 = 0.47 \text{ M x } 500 \text{ mL}$$

$$V_1 = 94 \text{ mL}$$

To make the 0.47 M solution, 94 mL of the 2.5 M HCl and 406 mL of H_2O will be added to a 500 mL volumetric flask.

In second part of the experiment, the amount of acetic acid in commercial vinegar will be determined. Acetic acid ($C_2H_4O_2$) is a weak organic acid found in vinegar. The structure is shown below. It will be reacted with the standardized sodium hydroxide (NaOH).

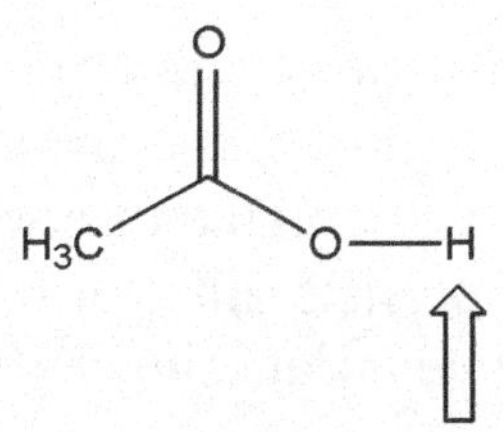

Acidic proton

The acid-base reaction:

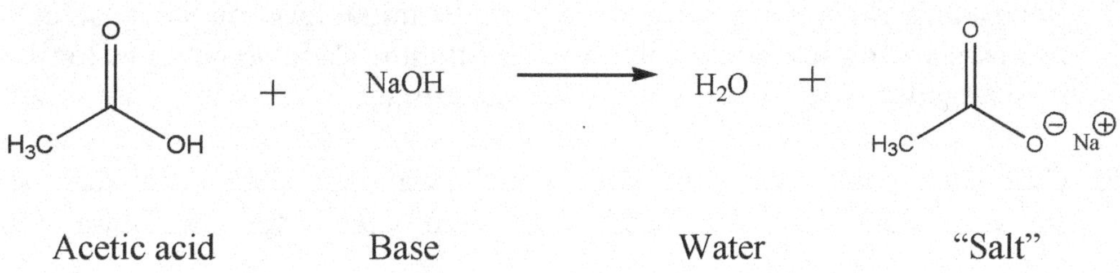

Acetic acid Base Water "Salt"

Equipment

Ring stand	Burette	250 mL Erlenmeyer flask	Volumetric flask
Stir bar	Magnetic stirrer	50 mL beaker	
Vinegar	NaOH	Graduated cylinder	

Experimental Procedure

Part 1 - Standardization of NaOH solution

1. Clean the burette by filling it with water and emptying several times.

2. Fill the burette with NaOH to the 0.00 mark on the top. Record the initial burette reading on your data sheet.

3. Measure out 0.50 g of PAT. Add to a 250 mL Erlenmeyer flask. Record the mass on your data sheet.

4. Add about 25 mL of distilled water and a magnetic stir bar to the flask. Place the flask on the stir plate. Turn on the stir plate to a medium setting until the PAT is fully dissolved.

5. Add 2 to 3 drops of phenolphthalein solution to the flask. The color should remain colorless. Have the solution gently stirring. Refer to Figure 11-1 for the setup.

6. SLOWLY titrate the NaOH solution into the flask by turning the stopcock parallel with the burette. Do not titrate past the 50.0 mark on the burette. Refill with NaOH if needed. Stop the titration once the end point is reached. This is when a pink color persists fully for more than 10 sec. Record the volume of NaOH used on your data sheet.

7. The contents of the flask may be disposed of down the drain. Rinse the flask and stir bar with water before using again.

8. Repeat this titration in triplicate. Record the initial level of NaOH solution from where you stopped in the previous titration. You do not have to fill the burette with 50 mL of NaOH solution each time.

9. Calculate the normality of NaOH for each titration, and then calculate the average NaOH normality. Refer to sample problem 3 for calculations.

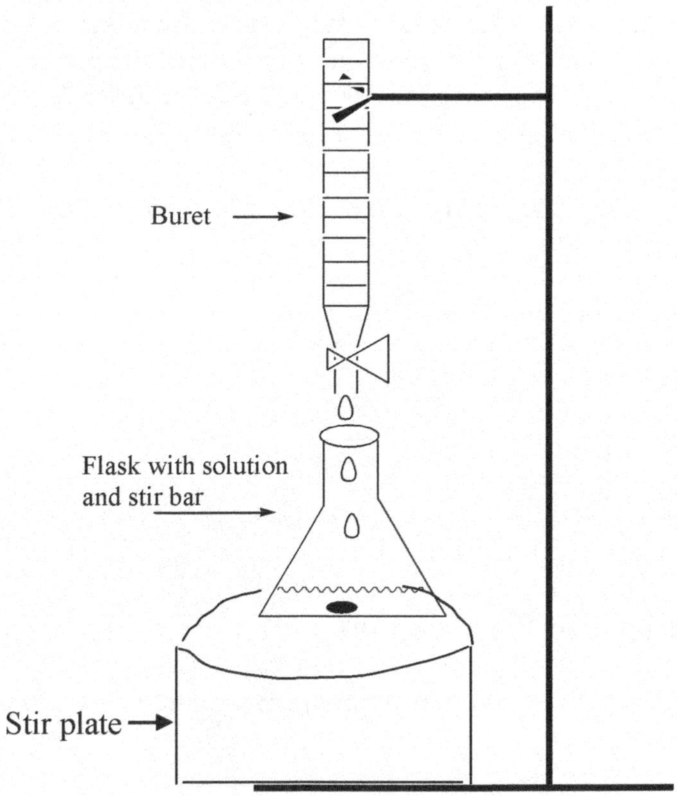

Figure 11-1. Titration Setup

Part 2 - Determination of percent acetic acid in commercial vinegar

1. Measure the mass of 4 g of vinegar into a 250 mL Erlenmeyer flask. Record the mass on your data sheet.

2. Dilute the vinegar with 25 mL of distilled water.

3. Add a few drops of phenolphthalein indicator solution to the flask. Also, add a magnetic stir bar into the flask.

4. Fill the burette with the standardized NaOH to the 0.00 mark. Place flask onto the stir plate under the burette.

5. Titrate with the standardized NaOH to the end point when the color of the solution remains pink for 10 sec. Do not titrate past the 50.0 mark on the burette. Refill with NaOH if needed. Record the amount of NaOH needed to reach the end point.

6. Repeat the titration in triplicate. Pour the waste down the drain. Rinse the flask and stir bar with water before reusing. Record the initial level of NaOH solution from where you stopped in the previous titration. You do not have to fill the burette with 50 mL of NaOH solution each time.

7. Calculate the percent acetic acid for each sample and then calculate the average percent.

Calculations

Part 1 - Standardization of NaOH solution

1. Divide the grams of PAT used by the equivalent weight of PAT to obtain the equivalents of PAT.

2. Divide the equivalents of PAT by L of NaOH used to obtain the normality of NaOH.

Part 2 - Determination of percent acetic acid in commercial vinegar

1. To find moles of base, multiply the molarity of base x liters of base used for titration.

2. To find moles of acid, moles of base = moles of acid from reaction equation. It is a one to one mole ratio.

3. To find g of acid, multiply the moles of acid by molar mass of acetic acid.

4. To calculate percent acetic acid, divide g of acetic acid by g of vinegar used, and then multiply by 100.

5. Calculate the average percent of acetic acid in commercial vinegar.

NAME: ______________________________

PARTNER: ___________________________

DATE: __________________

Titration of Acid and Bases

Part 1 - Standardization of NaOH solution

	Trial 1	Trial 2	Trial 3
1. Mass of PAT	______	______	______
2. Initial volume of NaOH	______	______	______
3. Final volume of NaOH	______	______	______
4. Volume of NaOH used (#3 - #2)	______	______	______
5. Liters of NaOH used (# 4 /1000)	______	______	______
6. Equivalents of PAT	______	______	______
7. Normality of NaOH	______	______	______
8. Average normality		____________	

SHOW CALCULATIONS.

Part 2 - Determination of percent acetic acid in commercial vinegar

Normality of standardized NaOH: ________ equivalents solute / L of solution
(from part 1 – Standardization)

Molarity of standardized NaOH: ________ mol/L

Molecular mass of acetic acid ($C_2H_4O_2$): ________ g/mol

	Trial 1	Trial 2	Trial 3
1. Mass of vinegar	______	______	______
2. Initial volume of NaOH	______	______	______
3. Final volume of NaOH	______	______	______
4. Volume of NaOH used (#3 minus # 2)	______	______	______
5. Moles of base	______	______	______
6. Moles of acid	______	______	______
7. Grams of acid	______	______	______
8. % Acetic acid	______	______	______
9. Average %		____________	

SHOW CALCULATIONS.

Questions

1. Determine the equivalent weights of the following acids or bases:
   a. KOH

   b. HF

   c. $Ca(OH)_2$

   d. H_2CO_3

2. Calculate the normality of NaOH if 48.9 mL of the solution is required to neutralize 3.15 g of PAT.

3. What would happen to the normality if you titrated too much NaOH into the flask? Explain your reasoning to whether the normality would decrease, increase, or stay the same.

Lab 12
Determination of the Amount of Citric Acid in Various Citrus Fruits

All fruits are acidic, to some extent; their pH values are less than 7. The acidity of citrus fruits, such as lemons, limes, and grapefruit, is primarily determined by the amount of citric acid in the fruit. What you will be doing today is to titrate a quantity of juice from a citrus fruit with sodium hydroxide, and so determine the percent citric acid in the fruit juice. We will assume here that all the acid in the fruit juice is citric acid, although there may be some other weak acids (such as ascorbic acid, vitamin C) present.

Citric acid has the following structure:

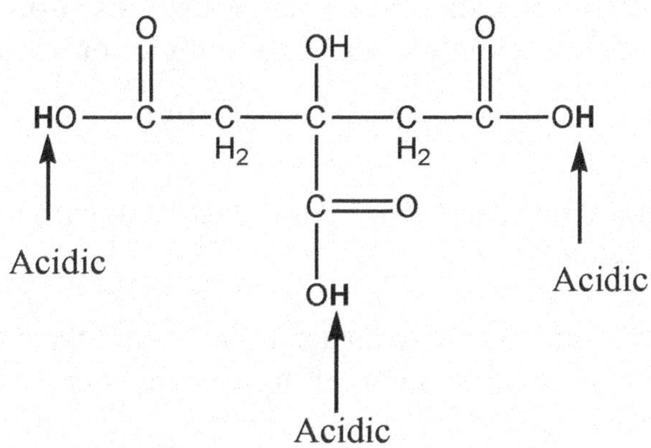

You'll notice that it is a triprotic acid, meaning that it has three hydrogens that can react with hydroxide in an acid-base reaction. These hydrogens are bold in the structure above.

Citric acid ($C_6H_8O_7$) reacts with sodium hydroxide in the following way:

$$C_3H_5O(CO_2H)_3 + 3NaOH \rightarrow 3Na^+ + C_3H_5O(CO_2)_3^{3-} + 3H_2O$$

Notice that, because it is a triprotic acid, it requires three moles of sodium hydroxide to react with (or neutralize) one mole of citric acid. Thus, once you have determined how many moles of sodium hydroxide it takes to neutralize the citric acid, the number of moles of citric acid you have will be one-third the number of moles of NaOH.

Equipment

Ring stand	Burette	125 mL Erlenmeyer flask	Stir bar
Magnetic stirrer	Fruit juice	Plastic pipets	
25 mL Graduated cylinder			

Experimental Procedure

1. Your laboratory instructor will squeeze the juice out of several citrus fruits: grapefruit, lemon, or lime. Choose one of the juices. Write its name on your data sheet.

2. Place an Erlenmeyer flask and tare it on a balance. Dispense using pipet about 1.5 g to 2.0 g of juice into flask. Record the exact mass of juice on your data sheet.

3. Add 20 mL of distilled water to the flask.

4. Add two drops of phenolphthalein to the Erlenmeyer flask with juice. The solution should be clear at this point.

5. Place Erlenmeyer flask under a burette on a stir plate. Add the magnetic stir bar to the flask. Turn on the stir plate to a medium setting. You do not want the solution to splash out of the Erlenmeyer flask.

6. Titrate the juice with ~ 0.1M NaOH (the exact concentration of the NaOH will be given to you). The end point will be when the color of the solution stays pink for 10 sec. Record exactly how many milliliters it took to reach the end point.

7. Perform the titration twice more with the juice.

8. Repeat the titration procedure using a second juice.

Calculations

1. To find moles of NaOH used, multiply L of NaOH used by the molarity of NaOH that was given. (#5 x molarity of NaOH)

2. To find moles of citric acid present, divide moles of NaOH used by 3. This takes into account that 3 moles of NaOH are required to neutralize 1 mole of citric acid. (#6 / 3)

3. To find grams of citric acid, multiply moles of citric acid by molecular mass of citric acid. (#7 x molecular mass of citric acid)

4. The percentage of citric acid present in the juice is found by dividing grams of citric acid present by the mass of juice used, then multiplied by 100. ([#8 / #1] x 100)

NAME: ______________________________

PARTNER: ___________________________

DATE: __________________

Determination of the Amount of Citric Acid in Various Citrus Fruits

Molarity of NaOH: _________ M   Molecular mass of citric acid: _________ g / mol

Juice # 1: __________________

	Trial 1	Trial 2	Trial 3
1. Mass of juice	________	________	________
2. Final burette reading	________	________	________
3. Initial burette reading	________	________	________
4. mL of NaOH used (#2 minus # 3)	________	________	________
5. Liters of NaOH used (#4 / 1000)	________	________	________
6. Moles of NaOH used	________	________	________
7. Moles of citric acid present	________	________	________
8. Grams of citric acid present	________	________	________
9. Percent citric acid in juice	________	________	________

10. Average percentage ________________

SHOW CALCULATIONS.

Juice # 2: _______________

	Trial 1	Trial 2	Trial 3
1. Mass of juice	________	________	________
2. Final burette reading	________	________	________
3. Initial burette reading	________	________	________
4. mL of NaOH used	________	________	________
5. Liters of NaOH used	________	________	________
6. Moles of NaOH used	________	________	________
7. Moles of citric acid present	________	________	________
8. Grams of citric acid present	________	________	________
9. Percent citric acid in juice	________	________	________
10. Average percentage		________________	

SHOW CALCULATIONS.

Questions

1. A titration was performed using NaOH and a fruit juice. Suppose 31.3 mL of 0.250 M NaOH was used to reach the end point. Calculate the percent citric acid if you started with 5 g of juice.

2. How would your percent citric acid vary if you titrated too much NaOH into the flask?

Lab 13
Determination of Unknown Sample Concentration Using Spectroscopy

When hydrogen and other gases are heated or excited electrically, they will emit light with characteristic frequencies. In contrast, matter will also absorb discrete frequencies of light. This process of absorption is exactly the opposite of emission. During the absorption of light, an electron undergoes a transition from a lower energy level (usually the lowest) to a higher energy level. (Figure 13-1) The electron gains energy in this process by absorbing a photon of light whose energy is given by the Einstein equation, $E = h\nu$. Planck's constant, h, is 6.626×10^{-34} J·s. The energy of the photon corresponds to the difference in energy between the higher and lower energy levels.

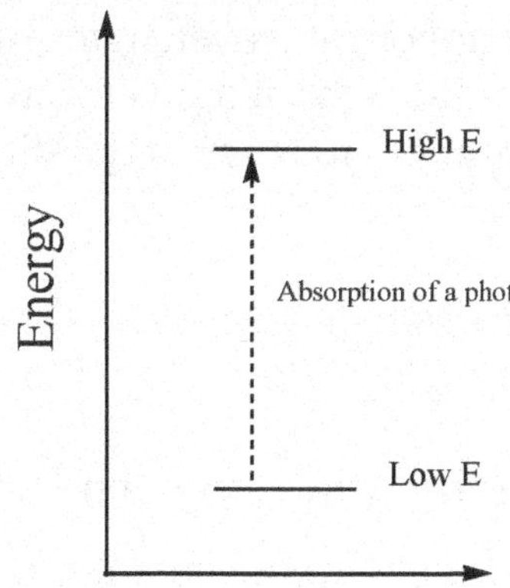

Figure 13-1. Energy level diagram

As a result of this energy gap, a component of light with a frequency ν is absorbed and other frequencies transmitted. Consequently, only one frequency will cause this transition. More often, a chemist refers to the wavelength of light, λ (lambda), rather than its frequency. Wavelength is equal to the speed of light, c, divided by the frequency. Einstein's equation then becomes:

$$E = \frac{h \times c}{\lambda}$$

When light of the correct wavelength shines through a solution of a substance that absorbs light, the intensity of the light will diminish as it passes through the solution. It diminishes because absorption will occur and this absorbance may be measured using the spectrophotometer.

The wavelengths of interest are the visible and near IR regions of the electromagnetic radiation spectrum. We will be measuring wavelengths from 380 to 800 nm.

Electromagnetic Spectrum:

Wavelength (m)

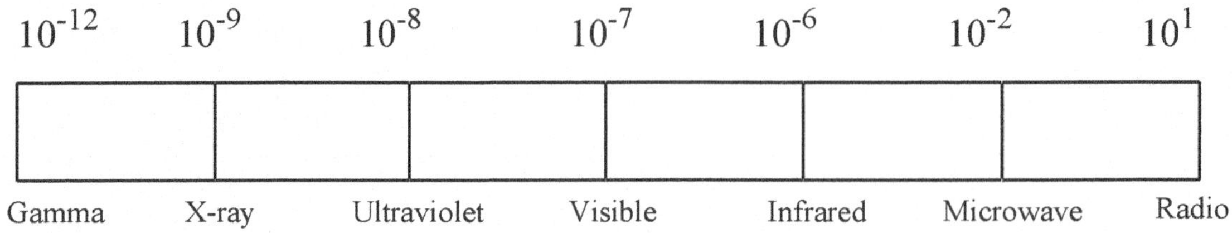

Spectrophotometer

Absorbance is measured with an instrument called spectrophotometer. The instrument separates light into its component wavelengths of visible color. It then selectively measures the intensity of the light of a given wavelength after it passes through a sample solution. All spectrophotometers, regardless of the manufacture, have certain common components.

1. A light source that produces light with a wavelength from about 375 nm to 650 nm, commonly tungsten or D_2.
2. A monochromator that selectively selects a particular wavelength of light sending it to the sample cell. Grating, prisms, or filters are used to select the desired wavelength.
3. A sample cell that holds the solution being analyzed. It is made out of quartz or glass depending on the wavelengths of interest.
4. A detector that measures the absorption (or transmittance) of the light. An electrical signal is produced when struck by photons. The signal is enhanced using either a photomultiplier tube or photodiode arrays.
5. A meter that indicates the absorption of light or transmittance of light.

Component setup for a spectrophotometer:

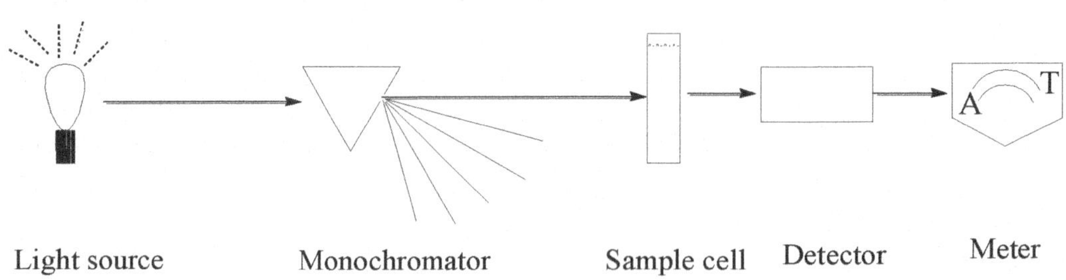

Transmittance is the fraction of light (P) that comes out of the sample cell on the other side. The equation is $T = P / P_0$, where P_0 is the light going into the sample. (Figure 13-2) The transmittance scale is from 0 to 1. Absorbance = - log T.

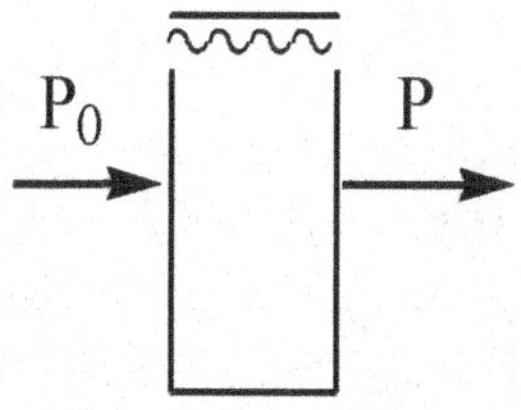

Figure 13-2.

Equipment

Spectrophotometer	2 - cuvettes	10 mL graduated cylinder	Kim wipes
Water bottle	Glass pipet	Blue rubber bulb	

Experimental Procedure

1. Turn on the spectrophotometer by rotating the power stitch knob clockwise and allow to warm up for about 5 minutes.

2. With no sample in the cuvette holder, turn the zero control knob (left knob) and bring the meter needle to 0 on the % transmittance scale (upper scale).

DO NOT TOUCH THIS KNOB AGAIN FOR THE REMAINDER OF THE EXPERIMENT.

Part 1 - Maximum wavelength determination

1. Turn the wavelength control knob to the desired wavelength beginning at 380 nm.
2. Fill the cuvette blank with distilled water and wipe the cuvette with a Kimwipe. Don't touch sides of cuvette after wiping. Insert into the sample holder. NOTE: Make sure (a) the cuvette is clean, (b) the cuvette is filled to the mark, (c) the cuvette has no air bubbles, (d) the cuvette is inserted properly, and (e) the cuvette holder cover is completely closed.
3. Rotate the Light Control knob (right knob) until the meter reads 0 on the percent absorbance scale (Lower scale). Remove the cuvette and save for later use.
4. Fill a second cuvette with the sample whose absorbance is to be measured. You will be using the full strength sample at 0.10 M. WITHOUT TOUCHING ANY CONTROLS, read percent absorbance and record on your data sheet to the corresponding wavelength.
5. Rotate the light control knob one full turn counterclockwise. Remove the cuvette containing the sample and set it in the holder.
6. Turn the wavelength control knob by increasing it from 380 nm to 400 nm. Go to step 2 to zero the instrument at the new wavelength. Move the wavelength up 20 nm each time.
7. Repeat the above procedure (steps 3 through 6) through 600 nm.
8. Plot a graph of absorbance vs. wavelength. Draw the best smooth curve between the plotted points.
9. Determine the maximum wavelength for your sample from your graph. Record the wavelength on your data sheet.

Part 2 - Dilution series

1. Set the spectrophotometer at the maximum wavelength determined from your graph of absorbance vs. wavelength using the wavelength control knob.
2. Fill the cuvette blank with distilled water and wipe the cuvette with a Kim wipe. Insert into the sample holder. NOTE: Make sure (a) the cuvette is clean, (b) the cuvette is filled to the mark, (c) the cuvette has no air bubbles, (d) the cuvette is inserted properly, and (e) the cuvette holder cover is completely closed.

3. Rotate the light control knob until the meter reads 0 on the percent absorbance scale (lower scale). Remove the cuvette and save it for later use.

4. Fill a second cuvette with the sample whose absorbance is to be measured. WITHOUT TOUCHING ANY CONTROLS, read the percent absorbance and record on your data sheet.

5. Empty the cuvette and fill with a new sample. For a new sample, accurately measure 5.0 mL of full strength sample, 0.100 M, into a10 mL graduated cylinder. Dilute the concentration in half by adding 5.0 mL of distilled water. This is now 0.0500 M, fill a cuvette with this solution. Measure the percent absorbance and record on your data sheet.

6. Accurately measure 5.0 mL of the 0.0500 M solution into the graduated cylinder. Dilute the concentration in half by adding exactly 5.0 mL of distilled water. This is now 0.0250 M, fill a cuvette with this solution. Measure the percent absorbance and record on your data sheet

7. Repeat the consecutive dilutions until 0.00625 M has been reached. Measure the percent absorbance with each dilution.

8. Plot a graph of absorbance vs. concentration. Make sure you draw the best straight line possible between the plotted points.

Part 3 - Unknown concentration determination

1. Obtain a solution of unknown concentration.

2. Zero the instrument with the blank of distilled water at the maximum wavelength.

3. Place the unknown sample in a cuvette. Wipe the outside of the cuvette with a Kim wipe.

4. Measure the unknown's percent absorbance using the procedure as above.

5. Graphically determine the unknown's concentration by interpolation.

NAME: ______________________________

PARTNER: ___________________________

DATE: __________________

Determination of Unknown Sample Concentration Using Spectroscopy

Part 1 - Maximum wavelength determination

Type of sample: ________________________

Wavelength (nm)	Absorbance
380	
400	
420	
440	
460	
480	
500	
520	
540	
560	
580	
600	

Maximum wavelength: ____________ nm

Part 2 - Dilution series

Concentration (M)	Absorbance
0.100	
0.0500	
0.0250	
0.0125	
0.00630	

Part 3 - Unknown concentration determination

Unknown absorbance: ______________

Unknown concentration: _____________ M

Questions

1. Why do you have to wipe the cuvette before taking the measurements?

2. Why do you line up the cuvette the same way each time?

3. From your second graph, what would the absorbance be if the concentration was 0.038 M?

Lab 14
Determination of the Molecular Mass of a Gas

The determination of the molecular mass of a volatile liquid will be obtained by using the Ideal Gas Law: PV = nRT. Substituting n = g / molecular mass (mm) into the Ideal Gas Law: PV = (g / mm)RT, and then solve the equation for molecular mass (mm).

$$\textbf{Molecular mass} = g \times R \times T / (P \times V)$$

A small sample of a liquid is placed in a pre-weighed flask and vaporized to expel all the air from the flask, leaving it filled with the vapor of a liquid at a known temperature and atmospheric pressure. The temperature will be the temperature of the boiling water bath. The flask plus the vapor is then cooled so that the vapor condenses. The mass of the flask plus condensed vapor plus air is then measured. The mass of the air, being nearly identical before and after, cancels out. The preceding data, in conjunction with the volume of the flask, permit the calculation of the molecular mass.

Equipment

Aluminum foil	125 mL Erlenmeyer flask	Rubber band	Pin
600 mL beaker	100 mL graduated cylinder	Hot plate	
Boiling chips			

Experimental Procedure

1. Record the number of your unknown liquid.

2. Place a small square of aluminum foil over the mouth of a clean, dry 125 mL Erlenmeyer flask. Fold the foil loosely around the neck of the flask and secure with a rubber band.

3. Make a very small hole in the foil using a pin. Record the mass of the flask, foil, and rubber band on your data sheet.

4. Carefully, remove the foil and rubber band. Place about 2 mL of the unknown liquid in the flask. Replace the foil and rubber band. (See setup top of next page.)

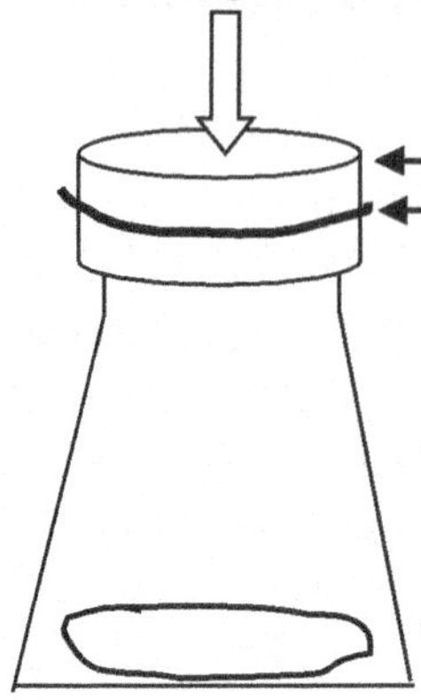

5. Immerse the flask (do not submerge) in a 600 mL Erlenmeyer flask that has been filled with tap water. Drop a few boiling chips in the water and heat to boiling using a hot plate. Time can be saved if the apparatus is set up at the beginning of the lab and the water is brought near a boil.

6. Once the liquid (and any condensed in the neck) has evaporated, record the temperature of the boiling water.

7. Record the barometric pressure from a barometer.

8. After the flask has cooled to room temperature and the vapor has condensed, wipe it dry and remove any water that may adhere to the aluminum foil. Measure the mass of the flask, foil, rubber band, and condensed unknown liquid.

9. Calculate the mass of the unknown liquid.

10. Remove the aluminum foil and fill the flask completely with water. Measure the volume by pouring the water into a 100 mL graduated cylinder. When measuring, you will have to measure and pour out the first 100 mL of water, then measure the remaining water in the flask. Record volume on your data sheet.

11. Calculate the molecular mass of the unknown liquid.

12. Repeat the experiment with the second unknown liquid.

1 atm

Lab 15
Preparation and Properties of Soap

Soap is the sodium or potassium salt of a long-chain fatty acid. Most solid soaps are sodium salts of the type to be made in this experiment. The starting materials are fats or oils, which are the glycerol esters of the fatty acids. A typical cooking oil, such as cottonseed or vegetable oil, and a concentrated solution of sodium hydroxide are the principal reactants. Ethyl alcohol is added to serve as a common solvent for the reactants and hence to speed up the reaction. It is not used in commercial soap-making operations.

The soap-making process reaction is called saponification. The reaction is shown below. The R groups are the nonpolar chains made of carbons and hydrogens of various lengths (12 to 18 carbons). They can contain double bonds giving unsaturated fats. Saturated fats have no double bonds in their nonpolar carbon chains.

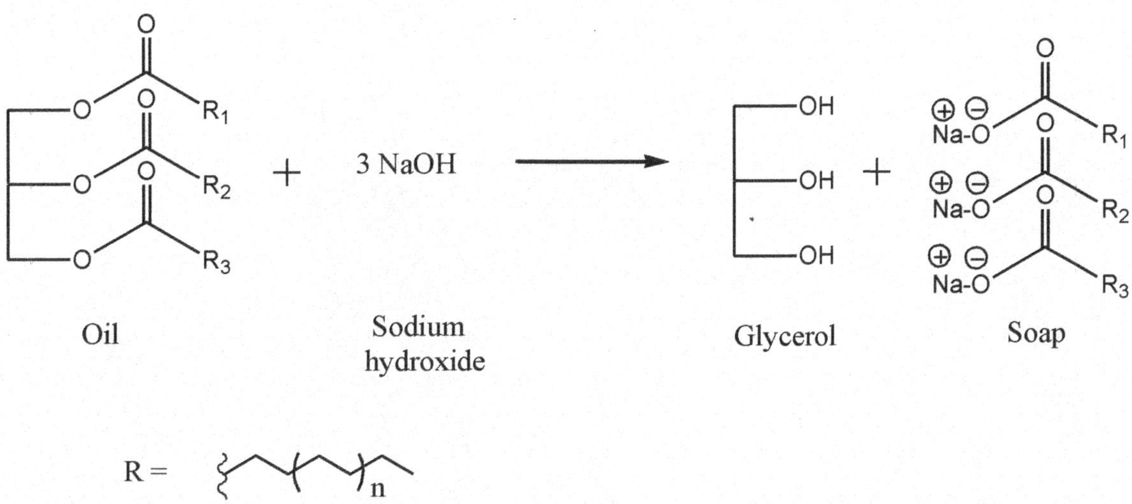

The sodium salt of the long-chain fatty acid is a typical surface-active agent. They have a polar end, which is hydrophilic (water-loving), and a long nonpolar chain, which is hydrophobic (water-hating). As a consequence, they can form emulsions by suspending oil in water when their nonpolar end group in towards the oil droplet and their polar end group out towards the water. (Figure 15-1) This is based on the idea that "like dissolves like", where similar compounds to the solvent can be dissolved. The oil droplet is similar in structure to the nonpolar chain of the soap molecule.

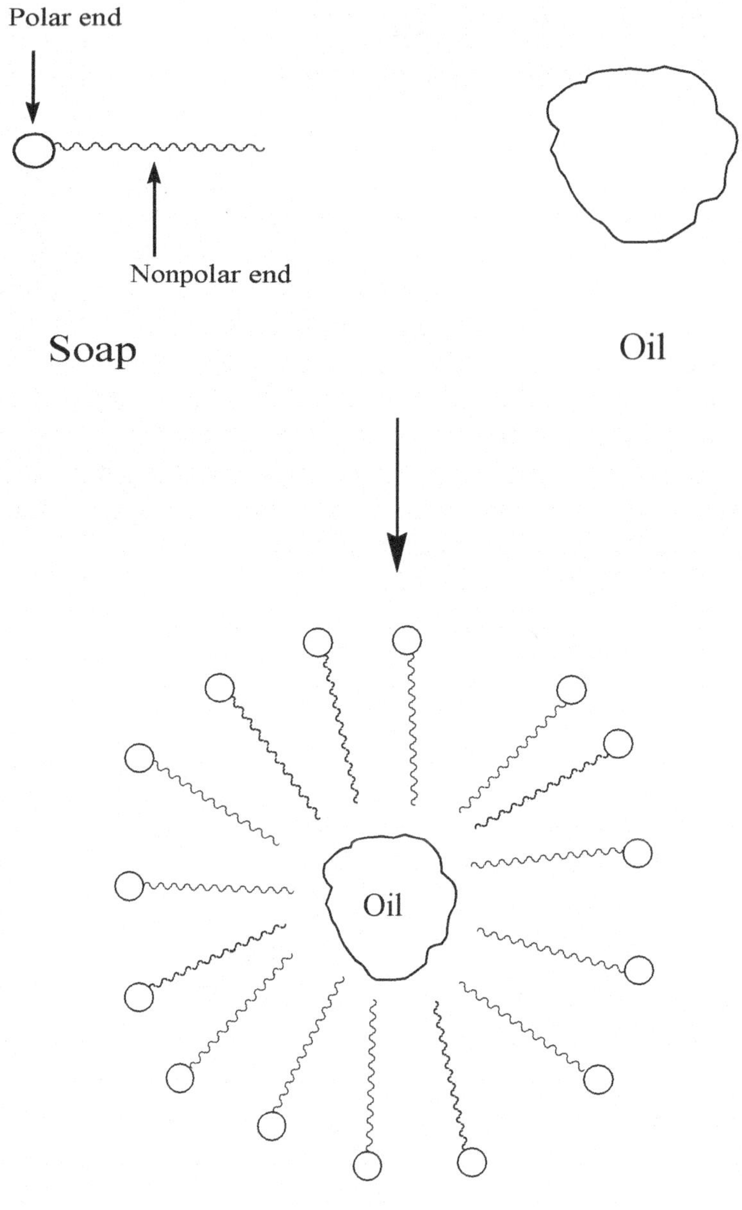

Figure 15-1. Emulsion formation of oil and water with soap.

When present, calcium, magnesium, iron, and other metal salts can form insoluble precipitates with the long-chain fatty acids. This kind of reaction is responsible for the problems that arise when soap is used with hard water. Hard water contains appreciable amounts of calcium, magnesium, and iron salts. Ion exchange filters are used in homes to help get rid of these ions to soften the water.

In this experiment, you will make soap from sodium hydroxide and vegetable oil. You will then test the washing properties, basicity, and effect of hard water using the soap.

Equipment

250 mL beaker	150 mL beaker	50 mL graduated cylinder
Stirring rod	Hot plate	Buchner funnel
Filter paper	4 - test tubes	100 mL graduated cylinder

Experimental Procedure

Part 1 - Synthesis of soap

1. Measure 10 mL of vegetable oil into a 150 mL beaker.
2. Add 10 mL of ethyl alcohol and 10 mL of 20 % sodium hydroxide (NaOH) to the beaker.
3. Stir the mixture thoroughly with a stirring rod.
4. Place the beaker on the hot plate. Turn on the hot plate.
5. Heat the solution gently, stirring constantly.
6. Continue to heat until a pasty mass remains in the beaker. There should be no odor of alcohol at this point. The pasty mass is a mixture of soap and glycerol.
7. Cool the beaker on the lab benchtop.
8. Add 100 mL of saturated NaCl solution to the soap mixture. Stir thoroughly with stirring rod. This process is called "salting out" and is used to remove the soap from water, glycerol, and any excess sodium hydroxide present.
9. Measure the mass of a piece of filter paper.
10. After mixing and stirring completely, place filter paper into funnel. Pour soap on top of filter paper. See the picture for setup. Turn on the water to start the filtering process. Once the dripping of water has ended, wash soap once with ice-cold water. The soap should be dry. Refer to Figure 15-2 for setup.

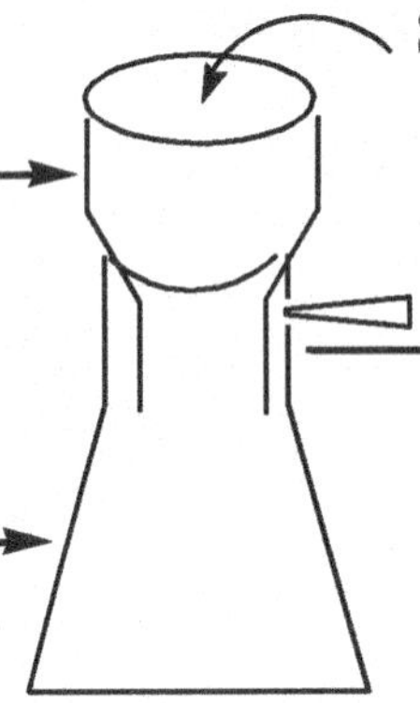

Figure 15-2. Funnel setup.

11. Measure the mass of the dried soap and record the mass on the data sheet.

Part 2 - Washing properties

1. Take a small amount of the soap and try to wash your hands with it. It should lather readily if soft water is used. If too much oil was used, the soap will feel greasy. If too much sodium hydroxide is still present, it will also make your hands feel slick, but will roughen your hands.

2. Record your observations on your data sheet.

Part 3 - Hard Water Reactions

1. Measure out 1 g of your soap and place in a 100 mL beaker.

2. Add 50 mL of water and warm using a Bunsen burner.

3. When you have obtained a solution, pour about 15 mL into each of three test tubes.

4. Test one of the three test tubes with 5 % $CaCl_2$ solution.

5. Add 5 % $MgCl_2$ solution to test tube #2 and add 5 % $FeCl_3$ solution to test tube #3.

6. Record your observations on your data sheet.

Part 4 - Basicity

1. Dissolve a small piece of soap in 5 mL of ethyl alcohol.

2. Add two drops of phenolphthalein indicator.

3. If the solution turns bright pink, it indicates the presence of NaOH.

NAME: ______________________________

PARTNER: ___________________________

DATE: __________________

Preparation and Properties of Soap

Part 1 - Synthesis of soap

Mass of filter paper ______________________

Mass of filter paper and soap ______________________

Mass of dry soap ______________________

Part 2 - Washing properties

Observations:

____

____

____

Part 3 - Hard water reactions

Observations:

1. $CaCl_2$:

____

____

2. $MgCl_2$:

3. $FeCl_3$:

Part 4 - Basicity

Observations:

Questions

1. How did the $CaCl_2$, $FeCl_3$, and $MgCl_2$ affect the lathering of the soap? Did you notice any residue on the sides of the test tube? If so, what is it commonly called?

2. What does the phenolphthalein test indicate?

3. What would have happened to your soap product if you didn't use a saturated NaCl solution for salting out?

Lab 16
Synthesis of Aspirin

Esters are derivatives of organic acids. They can be prepared by the reaction of carboxylic acids with alcohols in the presence of a catalytic amount of mineral acid. The following reaction serves to illustrate a reaction termed esterification.

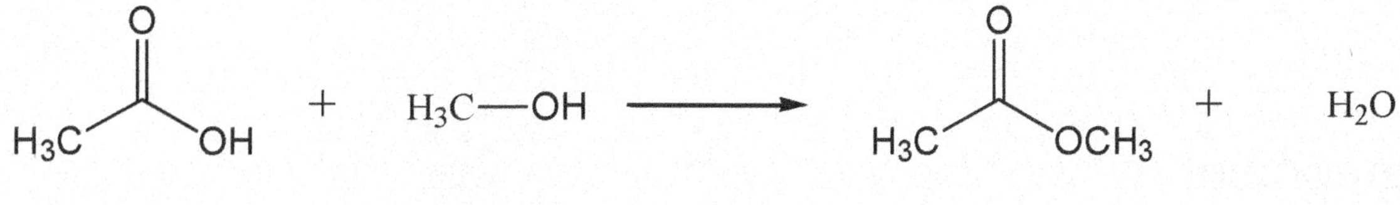

Acetic acid Methyl alcohol Methyl acetate

In this experiment, you will prepare an ester called acetylsalicylic acid. It is more commonly known as aspirin. Aspirin is one of the oldest and generally most useful drugs known. It is an analgesic (pain killer), antipyretic (reduces fever), anti-inflammatory, and most recently seems to reduce the risk of heart diseases. It can be prepared by heating salicylic acid and acetic acid for an extended period of time. However, instead of using acetic acid, you will use acetic anhydride because it is more reactive and will accomplish the esterification reaction more quickly. The lab will be done during two lab periods. The reaction is shown below.

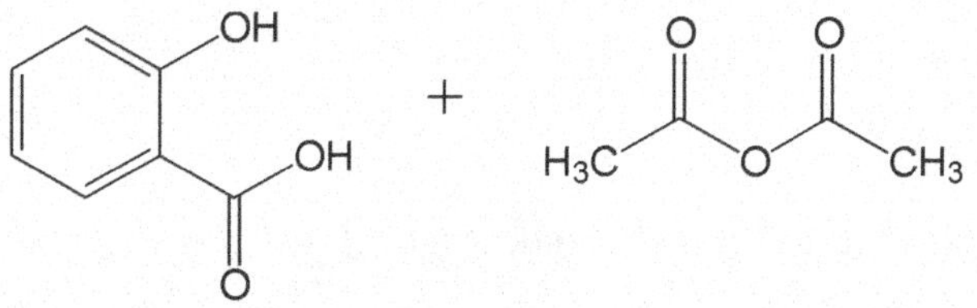

Salicylic acid Acetic anhydride

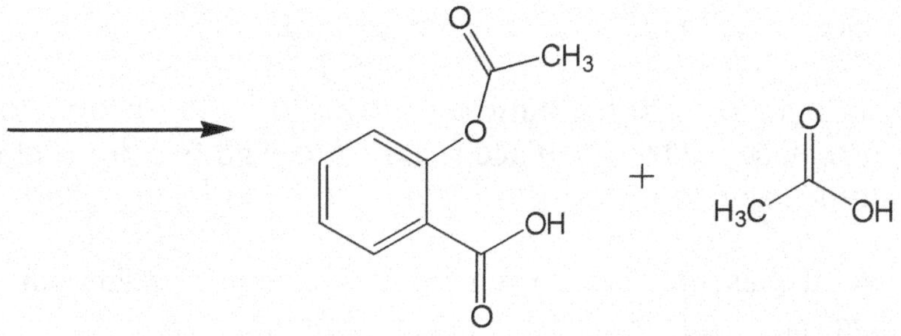

Acetylsalicylic acid

(Aspirin)

Acetic acid

The crude aspirin product will need to be purified. Most solids may be purified by recrystallization. This is usually achieved by dissolving the substance in a suitable solvent at the boiling point, filtering the solution by gravity to remove any suspended insoluble particles, and letting the crystallization proceed as the solution cools slowly.

As you may recall, the melting point is a physical property of a substance. The melting point will be used as a test of purity of a crystalline substance. Pure aspirin has a melting point of 135 °C. If a crystalline substance is impure, the melting point will be lower and the impure crystalline substance will melt over a larger range than pure material.

Equipment

125 mL Erlenmeyer flask	10 mL graduated cylinder	Hot plate	Ice
400 mL beaker	100 mL graduated cylinder	Buchner funnel	
4 -Test tubes	Capillary tubes	Filter paper	
Mini Bunsen burner	Ring stand	Thiel tube	

Experimental Procedure

Part 1 - Synthesis of aspirin

1. Set up a water bath by placing a 400 mL beaker filled ¾ with water onto hot plate. Begin heating to 80 °C to 90 °C.

2. Measure out 3.00 g of salicylic acid and record the mass on your data sheet. Place it into the 125 mL Erlenmeyer flask.

3. Once the water bath is between 80 -90 °C, CAUTIOUSLY and working in a fume hood, add 6 mL of acetic anhydride and five drops of concentrated sulfuric acid (H_2SO_4).

4. Mix the reagents by swirling the flask carefully. Clamp the flask to the iron ring stand. Lower the flask into a water bath.

5. Heat the flask in the water bath warmed to 80 – 90 °C for 10 minutes.

6. Remove the flask and allow to cool to room temperature.

7. Add 40 mL of distilled water and cool the mixture in an ice bath (while swirling) to complete the crystallization.
8. Filter using a Buchner funnel (as demonstrated by your instructor). Wash the crystals with a little ice-cold distilled water. Allow the crystals to air dry.

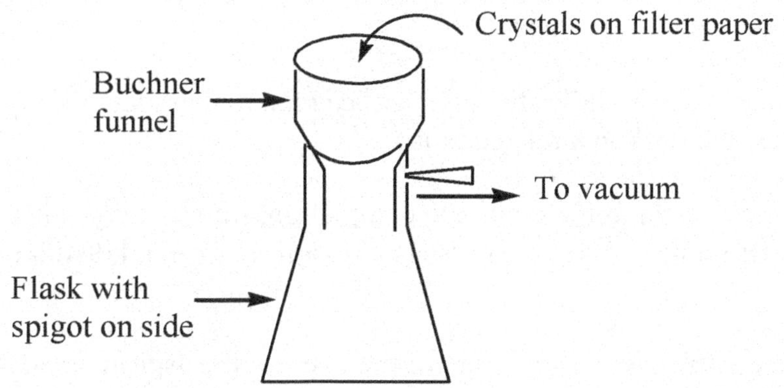

9. Measure the mass of the crude crystals.

Part 2 - Testing aspirin for purity

1. Dissolve a few of the crystals of salicylic acid in 5 mL of distilled water in a test tube.

2. Add one drop of 1 % ferric chloride solution. Note the color on your data sheet.

3. Dissolve a few of the crystals of your aspirin in 5 mL of distilled water in a test tube.

4. Add one drop of 1 % ferric chloride solution. Note the color on your data sheet.

Part 3 - Recrystallization of aspirin

1. Set up a water bath and heat.

2. Place your crude aspirin in about 10 mL of absolute ethanol in a 125 mL Erlenmeyer flask.

3. Warm the alcohol in a water bath until the aspirin crude crystals dissolve.

4. When the aspirin has completely dissolved, add 25 mL of warm (50 °C) distilled water. If any crystals form at this point, heat the solution in the water bath again until they dissolve.

5. Cool the mixture while stirring in an ice bath to complete the crystallization. or place the flask in a refrigerator.

6. After crystallization has come to completion, filter the crystals using the Buchner funnel. Wash the crystals with a little ice-cold distilled water, and air dry.

7. When completely dry, measure of the mass of the aspirin crystals.

8. Test your aspirin for purity as outlined in Part 2. Compare the color of the recrystallized aspirin with the crude aspirin and the salicylic acid.

Part 4 - Analysis of aspirin

1. Place a small amount of the purified aspirin in a capillary tube as demonstrated by your instructor.

2. Attach the capillary tube to a thermometer by means of a small rubber band cut from ¼ inch rubber tubing.

3. The thermometer and aspirin sample are placed in a Thiel tube containing mineral oil. Be sure that the sample and the thermometer tip are slightly below the bottom of the top side-arm neck of the tube.

4. Heat the side-arm of the tube using a Bunsen burner until the aspirin crystals in the capillary tube begin to collapse and melt.

5. Record the temperature at which the melting begins and the temperature at which the aspirin is completely melted.

6. Repeat with a second sample in a new tube.

NAME: ____________________

LAB PARTNER: ________________

DATE: ____________________

Synthesis of Aspirin

Part 1 - Synthesis of aspirin

Mass of Salicylic acid ______________

Mass of filter paper ______________

Mass of filter paper and crystals ______________

Mass of crude aspirin crystals ______________

Part 2 - Testing of aspirin for purity

Color of $FeCl_3$ with salicylic acid ______________

Color of $FeCl_3$ with crude aspirin ______________

Part 3 - Recrystallization of aspirin

Color of $FeCl_3$ with recrystallized aspirin ______________

Mass of filter paper ______________

Mass of filter paper and crystals ______________

Mass of recrystallized aspirin ______________

Part 4 - Analysis of aspirin

Melting Temperature:

Trial 1 ______________ Trial 2 ________________

Optional :

Yield of aspirin

Theoretical moles of aspirin ________________

Theoretical grams of aspirin ________________

% Yield of aspirin ________________

Molecular mass of aspirin = 180 g/mol

SHOW CALCULATIONS.

Questions

1. Why do you rinse the crystals with ice-cold water during the filtering process? Why not hot water?

2. If your yield was higher than 100 %, explain where during the experiment you could reduce the percentage.